社会科学文献出版社
SOCIAL SCIENCES ACADEMIC PRESS (CHINA)

# The Great Action

## In the New Era of Ecological Civilization

# 走向生态文明新时代的
# 大文化行动

2016生态文明贵阳国际论坛生态文化主题论坛讲演集

陈敏尔、雒树刚会见现场（一）
The meeting between Chen Min'er and Luo Shugang（1）

陈敏尔、雒树刚会见现场（二）
The meeting between Chen Min'er and Luo Shugang（2）

文化部部长雒树刚出席论坛
Luo Shugang, Minister of Culture of China at the Forum

雒树刚部长接见论坛嘉宾
Minister Luo Shugang greeting the Forum participants

时任中共贵州省委常委、省委宣传部部长张广智致欢迎辞
Zhang Guangzhi, Member of standing committee of Guizhou
provincial party and Propaganda Minister of Guizhou province
delivered a welcome speech

何力副省长与嘉宾亲切交流
Vice-governor He Li communicated with guests

何力副省长主持论坛
Vice-governor He Li hosted the Forum

文化部部长助理于群出席论坛
Yu Qun, Assistant Minister of Culture at the Forum

国家文物局副局长顾玉才出席论坛
Gu Yucai, deputy director of the State Administration of Cultural
Heritage of China at the Forum

文化部外联局局长谢金英出席论坛

Xie Jinying, Director-General, Bureau for External Cultural
Relations, Ministry of Culture of China, at the Forum

生态文化主题论坛开幕

The opening ceremony of Eco-Culture Forum

论坛现场

At the Forum

与会嘉宾

The guests

论坛现场

At the Forum

# 目 录

# CONTENTS

# 前　言

2016 年 7 月 9 日下午，由文化部、贵州省人民政府主办，贵州省文化厅、贵州大学、贵州省生态文明研究会承办的生态文化主题论坛在贵阳国际生态会议中心举行，论坛主题是"走向生态文明新时代的大文化行动"。

文化部部长雒树刚出席论坛并发表主旨演讲。时任中国宋庆龄基金会常务副主席齐鸣秋，联合国开发计划署驻华代表处国别主任文蔼洁，全球文化网络主席梅里·玛达沙西，中国社会科学院学部委员、民族文学研究所所长朝戈金，世界自然保护联盟环境法委员会副主席本·布尔，中国科学院地理科学与资源研究所研究员陈田，韩国首尔大学名誉教授全京秀，中国传媒大学教授卜希霆，贵州省文化厅厅长徐静等嘉宾发表演讲。

时任贵州省委常委、省委宣传部部长张广智在论坛上致欢迎辞，贵州省人民政府副省长何力主持论坛。文化部部长助理、办公厅主任于群，国家文物局副局长顾玉才，文化部外联局局长谢金英等出席论坛。来自文化部、中国宋庆龄基金会、中国传媒大学、中国人民对外友好协会、英国驻重庆总领事馆的嘉宾，中东

欧艺术家采风团，贵州省文化艺术界代表，媒体代表，共计 160
余人参加论坛。

现将部分嘉宾的演讲整理并摘要刊发，以飨读者。

# Preface

On July 9, 2016, the ecological culture forum was held in Guiyang international convention center. The forum was hosted by the Ministry of Culture of China and the people's Government of Guizhou province, undertaken by Cultural Department of Guizhou Province, Guizhou University and Guizhou Ecological Civilization Research Association. The forum's theme is "Big Action of Culture toward Ecological Civilization of The New Era".

Luo Shugang, Minister of Culture, attended the forum and delivered a keynote speech. Executive vice-president of China Soong Ching Ling Foundation Qi Mingqiu, country director of Representative Office in China of United Nations Development Porgram Agi Veres, chairwoman of the global culture network Mehri Madarshahi, member of Chinese Academy of Social Sciences and the head of Institute of Ethnic Literature Chao Gejin, vice-president of Environmental Law Committee of World Conservation Union Bernhard Willem Boer, the researcher of Institute of Geography Sciences and Natural Resources Research, CAS Chen Tian, honorary professor of University of Seoul in South Korea Chun Kyung Soo, professor of Communication University of China Bu Xiting and chief of Cultural Department of

Guizhou Province Xu Jing all delivered their speeches.

Member of standing committee of Guizhou provincial party and Propaganda Minister of Guizhou province Zhang Guangzhi delivered a welcome speech. The vice-governor of the people's Government of Guizhou province He Li hosted the forum. Assistant and office director of the Minister of Culture and Chief of External Relations Bureau Yu Qun, the vice-director of State Adminisration of Cultural Heritage Bureau Gu Yucai, the director of Bureau for External Cultural Relations, Ministry of Culture People's Republic of China Xie Jinying attended the forum. Over 160 people took part in the forum, and they are from the Ministry of Culture of China, China Soong Ching Ling Foundation, Communication University of China, Chinese People's Association for Friendship with Foreign Countries, England general consulate in Chongqing; Central and Eastern Europe artist group, representatives of Guizhou provincial culture and art circles and media representatives.

Here we particularly collated and published some guests' speeches for readers.

# 在生态文化主题论坛上的致辞

时任中共贵州省委常委、省委宣传部部长

张广智

**尊敬的树刚部长，各位来宾，女士们、先生们：**

　　贵山贵水迎贵客，文脉文化铸文明。作为生态文明贵阳国际论坛 2016 年会的重要议程——生态文化主题论坛今天正式开幕了。在这里，我谨代表贵州省委、省政府及主办单位对各位嘉宾的光临表示热烈的欢迎！

　　生态文明贵阳国际论坛作为中国唯一以生态文明为主题的国家级、国际性论坛，已成功举办了三届。随着该论坛知名度越来越高、影响力越来越大，我们的朋友圈也越来越广，绿色贵州的名声也越来越响。可以毫不夸张地说，生态文明贵阳国际论坛已成为贵州对外开放交流的一个重要的窗口、一张靓丽的名片。生态文明贵阳国际论坛为什么会落户贵州？主要是因为贵州人始终秉持"道法自然、美美与共"的生活理念和"天人合一、知行合一"的人文精神，这种理念和精神具有深厚的生态文化底蕴，也融入了贵州人的血脉之中。千百年来，贵州各族人民与大自然和谐共处，始终与万类生灵为邻，与树木花草为伴，创造了绚丽多彩的原生态文化，这可以说是生态文化的本源。到 20 世纪 80 年代，面对人为造成的生态危机，贵州人民开启了以扶贫开发、生态建设为主要内容的毕节试验区建设，取得了重大成果，积累了丰富经验。特别是近年来，我们深入贯彻落实习近平总书记对贵州的重要指示精神，唱响"多彩贵州拒绝污染"的时代强音，坚持生态优先、绿色发展、"两线"一起守、"两山"一起建，加快推进生态文明先行示范区和山地公园省建设，在生态文化发展和生态经济建设方面进行了积极探索和艰苦努力，取得了令世人瞩目的成绩。可以说，生态论坛因天时、地利、人和而生，生态文化因绿色、纯朴、共享而兴，生态贵州因山美、水美、人美而多彩。

建设生态文明，离不开经济社会发展和生态文化环境。我们正在走向生态文明新时代，积极践行"绿色发展·知行合一"的施政理念，树立起了生态文化的高度自觉和自信。从这个意义上讲，我们本次主题论坛聚焦生态文化这一主题开展深入交流探讨，是顺势而为、应势而彰。我们衷心期待这次主题论坛取得丰硕成果，也热切期望这个论坛持续办下去，越办越精彩。

最后，预祝生态文化论坛取得圆满成功！

# Address on the Ecological Culture Forum

Zhang Guangzhi

Distinguished Shugang Minister, ladies and gentlemen, fellow guests,

Good morning, on behalf of Guizhou provincial committee and government and host organizations, please allow me to extend our warmest welcome to all of you. As an important agenda of the Eco Forum Global Annual Conference Guiyang 2016, Ecological culture forum formally opened today.

Eco Forum Global Annual Conference Guiyang, as the only state-level international eco forum in China, has already been held three times successfully. With the increasing influence of the forum, the circle of friends of Guihzou is now getting wider and wider and green Guizhou is getting louder and louder. It has already been an important window for Guizhou international cultural exchange. Sticking to the firm belief of "naturalness of Taoism" and "harmony of human and nature" contributed to the coming of Eco Forum Global Annual Conference Guiyang, people of Guizhou have created a splendid ethnic culture by harmonious coexistence between man and nature through the centuries based on their rich cultural reserves. In the 1980s, when ecological crisis became a global challenge, we made great efforts for the construction of Bijie ecological demonstration zone by launching poverty alleviation and ecological improvement. Especially in recent years, we thoroughly implemented the spirit of the General Secretary Xi's important speeches, uttered the strongest voice of the times "colorful Guizhou without pollution". We continued to give high priority to ecology and green development to promote the constructions of ecological demonstration zones and mountain parks And we have obtained compelling achievement in ecological culture and ecological economy development after tireless efforts and arduous exploration. All in all, the Eco Forum comes at the right time, right place and with the support of people; ecological culture thrives for the green development, purity and co-sharing; Guizhou becomes colorful for the green mountain, clean water and kind people.

In the new era, ecological civilization cannot get away from economic development and ecological culture environment. We should actively practice the concept of "green development, unity of knowledge and action" to establish high awareness and confidence of ecological culture. In this sense, the Eco Forum comes as a result of trend of times. We hope that this forum will yield great results with our joint efforts, and become better and better.

At last, I wish the forum a great success! Thank you!

# 生态文明贵阳国际论坛 2016 年年会
## Eco Forum Global Annual Conference Guiyang 2016

Speakers
演讲嘉宾

雒树刚

Luo Shugang

中华人民共和国文化部党组书记、部长
Minister of Culture, the People's Republic of China

　　雒树刚，河北南宫人，中央党校理论部党的学说和党的建设专业毕业，研究生学历，硕士学位，编审。

　　1978 年 10 月至 1982 年 7 月在中国人民大学科学社会主义系学习，1983 年 9 月至 1986 年 7 月在中共中央党校理论部党建专业学习。历任《求是》杂志社政治理论部主任，中共中央宣传部政策法规研究室主任、副秘书长兼理论局局长。2000 年 3 月任中共中央宣传部副部长，2008 年 5 月任中共中央宣传部常务副部长兼中央精神文明建设指导委员会办公室主任。2014 年 12 月任文化部党组书记、部长。

# 以文化为舟
# 驶向生态文明新时代

——在 2016 年生态文明贵阳国际论坛生态文化主题论坛上的主旨演讲

### 中华人民共和国文化部部长　雒树刚

文化部部长雒树刚做主旨演讲

## 女士们、先生们、朋友们：

很高兴能够出席 2016 年生态文明贵阳国际论坛——文化主题论坛，同大家一起探讨文化在人类社会步入生态文明的进程中所应发挥的独特作用。2013 年，习近平主席在致生态文明贵阳国际论坛年会的贺信中指出，"保护生态环境，应对气候变化，维护能源资源安全，是全球面临的共同挑战"。面临这个共同挑战，全球有识之士纷纷从政治、经济、社会、科技等各个领域提出了众多的应对策略和解决方案。但由于生态环境保护的复杂性、长期性，解决这个问题不光要从体制、技术层面深入思考，更要从理念、文化等方面入手，树立尊重自然、顺应自然、保护自然的生态文明理念，形成天人合一、天人互益的生态文化，从根本上为生态环境保护的长期开展奠定基础。从这个意义上讲，文化是生态环境保护的重要内核。因此，我们这个文化主题论坛的召开可谓恰逢其时。我今天的演讲将首先简要分析生态文明建设的总体形势，继而和大家分享我关于文化在生态文明建设中的重要作用以及如何运用文化推动绿色发展的一些思考。

## 一　生态文明是人类社会发展大势所趋

生态环境是人类生活的物质基础，是永续发展的必要条件。人类社会先后经历了原始文明、农业文明和工业文明，在这个发展过程中，人类对自然的探索开发力度不断加大，对自然环境的破坏也随之加剧，并在工业文明时代达到了顶峰。尽管人类自工业革命以来创造了前所未有的经济奇迹，积累了巨大物质财富，但正如恩格斯在《自然辩证法》中所言："我们不要过分陶醉于我们人类对自然界的胜利。对于每一次这样的胜利，自然界都对我们进行报复。"人类逐步意识到，失去了良好的生态环境，一切经济社会发展都无从谈起，一切人类社会进步都将是镜花水月、转瞬即逝。

中华文明历来强调天人合一、尊重自然。中国政府也高度重视生态文明建设。中国共产党第十八次全国代表大会把生态文明建设纳入中国特色社会主义事业五位一体总体布局，明确提出大力推进生态文明建设，努力建设美丽中国，实现中华民族永续发展。在 2016 年 3 月公布的《中国国民经济和社会发展第十三个五年规划纲要》中，将生态文明建设作为重要组成部分，提出要落实创新、协调、绿色、开放、共享的发展理念，通过科技创新和体制机制创新，实施优化产业结构、构建低碳能源体系、发展绿色建筑和低碳交通、建立全国碳排放交易市场等一系列政策措施，形成人与自然和谐发展的格局。

## 二　生态文化是生态文明建设的重要基石

我刚才提到了生态文明是大势所趋，但趋势能否早日转化为现实则取决于全球各国和各国人民是否有决心和努力。而要实现人人为环保添砖加瓦、人人为生态环境尽心尽力的局面，首先需要大力普及生态文化，树立与生态文明相适应的新理念。生态文化是生态文明建设的重要基础，只有人人自觉自愿、积极主动地投入生态文明建设这一与每个人都息息相关的伟大事业中来，有利于生态保护的制度才能冲破层层现实利益的阻隔得以实现，有利于生态保护的技术才能在点点滴滴创新细流的汇聚下成汪洋之势、实现重大突破。总之，生态文化蔚然成风是生态文明的重要前提和必要基础。

在普及生态文化、树立生态新理念的过程中，我们要充分发挥文化的教育功能。人与自然的关系是艺术创作的重要主题和素材，许多艺术家在与自然和谐共处的过程中获得了灵感。我们要不断鼓励、引导以生态保护为主题，表现绿色发展理念的艺术作品的创作和传播，用优秀的文艺作品熏陶、引导大众，使环保理念润物无声、随风入夜般地进入各国人民心中，使各国人民更加自觉地珍爱自然、更加积极地保护生态。

对于生态文明题材的文艺创作而言，中华优秀传统文化是取之不尽、用之不竭的无穷之藏。崇尚自然、追求人与自然和谐相处进而融为一体是中华优秀传统文化的基因，当代文艺创作要善于从这个无比巨大的文化资源库中寻找灵感，通过创造性转化和创新性发展，将传统文化基因和当代艺术手法有机结合，创作出更多直达人心的生态主题文艺作品。

### 三　践行绿色发展，以文化之力推动生态文明建设

本次国际论坛的主题是"走向生态文明新时代：绿色发展，知行合一"。文化在"知"和"行"两方面都具有独特作用。在"知"这方面，我刚才已经谈了文化在推动生态文明理念形成和普及方面的启迪、引导作用。在"行"这方面，文化同样能够发挥巨大作用。

很多人认为，生态环境保护和生产力发展是一对矛盾。在工业文明时代，良好的生态环境和生产力也许就像鱼和熊掌，不可兼得。但我相信我们有智慧、有能力调和这对矛盾。《中国国民经济和社会发展第十三个五年规划纲要》将绿色发展作为五大新发展理念之一。而绿色发展理念要求我们将生态保护和生产力提高统一起来，树立起保护生态环境就是保护生产力、改善生态环境就是发展生产力的新理念。绿色发展的前提和基础是要推动理念和实践的转型升级，形成节约资源和保护环境的空间格局、产业结构、生产方式和生活方式。而文化在这一转变的过程中大有可为。

首先，绿色的空间格局需要文化。国土是生态文明建设的空间载体。生态文明建设要取得成功，必须要科学布局生产空间、生活空间和生态空间。而文化是打造绿色空间必不可少的重要元素。无论是保护具有历史文脉的古城、古镇、古村落，还是因地制宜地建设各具特色和文化底蕴的现代化城市和美丽乡村，都需要文化发挥潜移默化的影响作用。文化就像是建筑空间的灵魂，只有文化发挥了点睛之笔的作用，建筑空间才能真正成为流动的音乐，城市才能避免千城一面，做到各领风骚。

其次，绿色的产业结构和生产方式需要文化。文化产业是真正的绿色产业，文化产业增加值的不断提升是绿色产业结构形成的重要基础。在"十三五"期间，我们将大力推动文化产业发展，使文化产业增加值占 GDP 比重达到 5%，成为国民经济支柱性产业。可以想见，当动漫、游戏、创意设计、网络文化、数字文化服务等新型文化业态大发展大繁荣之时，也是绿色产业结构逐步成型之刻。此外，文化产业在自身不断发展壮大的同时，还将与制造、建筑、信息、旅游、农业、体育等相关产业融合发展，推动其他产业向着低碳、可持续的绿色方向发展。当文化作为生产力的价值不断被发掘时，绿水青山和金山银山将不再是非此即彼的选择，而是两者兼得的动态平衡。

最后，绿色的生活方式需要文化。我刚才提到，文化可以推动生态理念的形成，影响大家的行为，帮助大家自觉践行绿色生活方式。除此之外，对于文化产品和服务的消费本身就是一种绿色生活方式。今后一段时期，我们将从供需两端发力，着力扩大有效文化供给，改善文化消费条件，推动建立扩大和引导文化消费的长效机制。同时，为了保障人民群众的基本文化需求，使无力进行文化消费的群体也能获得基本的文化享受，我们要大力完善现代公共文化服务体系，全面推进基本公共文化服务的标准化、均等化，使文化产品和服务的消费成为大众生活方式的有机组成部分，推动绿色生活方式的逐步成型。

女士们、先生们、朋友们！

应对气候变化和保护生态环境是人类共同的事业，是人类命运共同体建设的重要组成部分，需要世界各国同舟共济、携手同行。生态文明贵阳国际论坛举办至今，已发展成为聆听各国声音、凝聚全球智慧，推动生态文明建设的重要国际平台。我希望，这个论坛能够越办越好，凝聚更多智慧、汇聚更大力量，为人类生态文明建设不断做出新的更大贡献。

谢谢大家！

# Culture as a Vehicle to Usher in the New Era of Ecological Progress

## Address Delivered at the Eco Forum Global Annual Conference 2016

### Luo Shugang, Minister of Culture

## Personal resume

Mr. Luo Shugang was born in Nangong, Hebei Province. He holds a Master's degree in party theories and capacity-building of CPC, and is an accredited editor.

Since December, 2014

Minister of Culture of China

From May, 2008-December, 2014

Executive Deputy Minister, Publicity Department, CPC Central Committee of China and concurrently Head of the General Office, Central Commission for Guiding Cultural and Ethical Progress of China.

From March, 2000 to May, 2008

Vice Minister, Publicity Department, CPC Central Committee of China.

Before March, 2000

Successively served as the Director of the Department of Political Theories, Qiushi Magazine, the Director for Policy and Legislation Research, Vice Secretary-General and concurrently Director-General of the Bureau for Political Theories, Publicity Department, CPC Central Committee of China.

From September, 1983 to July, 1986

Student of the Party School of the Central Committee of the Communist Party of China.

From October, 1978 to July, 1982

Student of Scientific Socialism, Renmin University of China.

Ladies and gentlemen, dear friends,

It is a great pleasure to attend this forum and join you in the discussion about the role of culture in the ecological progress of humanity. Chinese President Xi Jinping pointed out that "protecting the environment, addressing climate change and securing energy and resources is a common challenge for the whole world" in his message to the Eco Forum Global Annual Conference Guiyang 2013. In face of this challenge, the visionary people from all across the globe have suggested various solutions and strategies from political, economic, social and technological approaches. However, ecological environment protection is a long-term and sophisticated progress that requires not only institutional and technical support, but also the awareness to respect, protect and accommodate ourselves to the needs of nature, which provides a solid foundation for the sustainable preservation of the ecosystem and environment. In this sense, culture is an important core in ecological conservation and environment protection. At this timely and momentous forum, I would like to firstly outline the general situations of our ecological progress, and then share with you some of my thoughts on the role of culture in this progress and in facilitating green development.

First, ecological progress is the development trend of the human society.

Eco system and environment provides material foundation for the livelihood and sustainable development of the people. In the primitive, agricultural and industrial societies, human beings have explored and exploited the nature resures so much that great damage has been aggravated, especially in the industrial era. Indeed, the industrial revolution ushered in unprecedented economic miracles and created a huge amount of wealth, but as Frederick Engels stated in his *Dialectics of Nature*: "Let us not, however, flatter ourselves overmuch on account of our human victories over nature. For each such victory nature takes its revenge on us." Later, we became increasingly aware that without a well-preserved environment, economic and social achievements would be hollow and in vain, so would be any reference to human progress.

Since ancient times, the Chinese culture has valued harmony between man and nature and emphasized on respecting nature. At present, the Chinese government also attaches high importance to ecological progress. The 18th National Congress of the CPC listed ecological progress along with economic, political, cultural and social progress as the five goals in the overall plan for the cause of Chinese socialism, vowing to promote ecological progress to build a beautiful China and achieve lasting and sustainable development of the Chinese nation. The 13th Five Year Plan of China published in March, 2016 stressed the need to promote innovative, coordinated, green, open, and shared development, with ecological progress high up in the agenda. Technological and institutional innovation and reform should be encouraged to upgrade industrial structure, and policies should be implemented to build a low carbon energy system, develop green buildings and low carbon transportation, and foster a national carbon emission trading market to strike a balance between man and nature in our development.

Second, ecological culture is an important cornerstone of ecological process.

As I said earlier, ecological progress is an irresistible trend with which we all shall go along. The more we are committed to strive for it, the sooner we will make it a reality. To pool everyone's strength to protect the environment, an eco-friendly culture and public awareness of the importance of environment conservation should firstly be fostered. The eco-friendly institutions and measures can only be implemented despite their threat to some vested benefits, and greentech can only achieve breakthroughs with the conscientious and proactive engagement and dedication of every individual who acknowledges the relevance of ecological progress to their own life. Therefore, an eco-friendly culture becomes an important precondition and necessary foundation for ecological progress.

Culture nourishes and shapes our minds, and it can also help raise our ecological awareness. As artists are always inspired by their contact with nature, many artistic creations tell stories about man and nature. We should encourage and commission the production and promotion of outstanding artworks with the theme of environment conservation and green development. Such artworks can reach the heart of the people in a subtle and natural way, and raise the awareness of the need to protect and value the nature and ecosystem.

It is in the genes of the Chinese traditional culture that the nature should be respected, and man and nature should co-exist in harmony. The rich cultural wealth and traditions of China, therefore, can provide inexhaustible source materials for arts creation with ecological themes. We encourage contemporary artists to apply their creativity and innovative spirits to the integration of traditional culture and modern techniques, and produce artworks that are more compatible with modern times and more appealing to the public.

Third, we should fulfill green development and power ecological progress with culture.

The theme of the forum this year is Towards a New Era of Ecological Progress: Learning by Doing to Promote Green Development. Culture applies to both learning and doing in a unique way. For the learning part, I have talked about the role of culture in raising ecological awareness. Now I would like to talk about how culture can facilitate the doing, or practices, in promoting ecological progress.

Many people believe that productivity has to be developed at the cost of environment. It was probably true in the industrial society, but we believe that today, we are wise and capable enough to accommodate productivity with the needs of nature. The idea of green development, stated as China's 13th Five Year Plan, underlines the integration of productivity enhancement and ecological progress. It advocates the notion that by conserving the environment, productivity will also be fostered, and by improving the environment, productivity will also be advanced. The precondition and footing for green development is the upgrading and restructuring of strategies and practices which will usher in resource efficiency and eco-friendly industrial structure, production mode and lifestyle. The role of culture can be significant in such a restructuring process.

Firstly, an eco-friendly structure for space use asks for cultural insights. It is through land use that ecological progress can be advanced, and such progress requires us to allot space to production, to daily life and to ecological development as appropriate. Culture is an integral part in a green space structure. Its subtle influence matters in the preservation of historic cities, towns and villages, as well as in building culturally enriched and distinctive urban and rural communities. Culture places soul into architecture, and only when cultural touch is added, architecture can become lively enriched, and the cities can boast their unique charm, without worrying about being dully identical with one another.

Secondly, green industrial structure and production mode requires cultural involvement. Cultural industry is by all means green industry, and its increasing added value actually forges the pedestal for green industry structure. During the 13th Five Year Plan period, we will spare no efforts to energize cultural industry, increasing the proportion of its added value in GDP to 5%, so as to make it a pillar industry for national economy. The structure and framework of green industry will be substantial when new cultural businesses such as cartoon and animation, game, creative design, online culture, and digital cultural services embrace remarkable growth and prosperity. Moreover, we believe that cultural industry should be integrated with related industries including manufacturing, architecture, information, tourism, agriculture and sports to find for them a more low-carbon and sustainable development mode. With culture serving as an important factor that contributes to productivity, it could strike a balance between ecological and economic development and generate momentum for both.

Moreover, culture helps foster a green lifestyle. I mentioned that culture can raise people's ecological awareness, influence their actions and encourage people to live a more eco-friendly life. Apart from that, living a green life also means to spend more money on green products like cultural products and services. In the coming years, we will endeavor to increase efficient supply of cultural products and services, give more and easier access to the people to spend on culture, and foster a long-term mechanism that encourages more and quality cultural spending. In addition, we will make more efforts to improve our modern public cultural service system and promote equal access to standardized public cultural services in an all-round way to guarantee the basic cultural rights of the people and provide basic cultural enrichment to those who cannot afford to spend on culture. We hope to make it an integral part of people's life to spend on cultural products and services so as to pursue a greener lifestyle.

Ladies and gentlemen, dear friends,

It is a shared mission of the entire humanity to cope with climate change and protect the environment. It is so important to our community of common destiny that we should all join hands and support each other for this cause. Today, the Guiyang Eco Forum Global has become an important international platform to hear voices, share wisdom and pool strengths to promote ecological process. I wish the Forum greater success this year and in the future, and hope it can make continuous contributions to the ecological process of the world. Thank you.

生态文明贵阳国际论坛 2016 年年会
Eco Forum Global Annual Conference Guiyang 2016

Speakers

演讲嘉宾

文蔼洁

Ms. Agi Veres

联合国开发计划署驻华代表处国别主任

The Country Director of the United Nations Development

Programme (UNDP) in China

　　文蔼洁女士现任联合国开发计划署驻华代表处国别主任，负责管理开发署驻华代表处的项目及运行、战略统一以及合作伙伴关系。

　　文蔼洁女士自 2002 年起先后在联合国开发计划署担任多项职务。自 2011 年 10 月至 2014 年 8 月，她担任联合国开发计划署驻莱索托副代表；2009 年至 2011 年，她作为高级项目协调员在独联体、欧洲布拉迪斯拉发区域中心工作；2006 至 2008 年，在布拉迪斯拉发区域中心担任政策支持与项目发展副主任；2005 年至 2006 年，在开发署纽约总部担任多项职务，包括报告协调员和企业分析组组长；2002 年至 2005 年担任信息与通信技术传播官。在加入开发署之前，文蔼洁女士在埃森哲匈牙利、纽约分公司从事组织改革管理咨询，专注领域为组织有效性、传播与流程重组。

# 可持续发展下的文化变革

文蔼洁

联合国开发计划署驻华代表处国别主任文蔼洁女士演讲

非常感谢！女士们、先生们，很荣幸能参加这次生态文明论坛。正如刚才讲话中提到的，这次贵阳生态文明论坛第一次将文化因素包含进来，生态文明和可持续发展联系密不可分，二者有很多相似之处。

按照生态的足迹和其发展规律，我们应该正视当前面临的极限和挑战，如果继续现在的生活方式和消费习惯，据相关研究，到 2030 年我们的人口总量可能激增到要两个地球才够容纳，然而在今后的 15 年内我们不可能找到第二个类似的星球。避免这种情况的最好方法，就是改变当前的生产和消费文化。纵观人类历史的发展，我们过度使用地球资源，会给贫困和弱势群体带来负面影响，甚至会影响安全和发展问题的走向。

所以从文化的角度，我今天想要谈的是，如果按照今天的方式来继续生活、生产和消费，会有什么样的后果呢？文化和行为方式的转变，是我们唯一实现可持续发展未来的方法，以确保我们的后代不会因为我们现在对资源的过度攫取、破坏生态适应性而遭受负面影响。在论述文化的重要性时，我想引用阿玛提·山姆的一句话来谈谈文化和生态文明之间的关系，文化是生活中无形的一部分，如果发展可以提高我们的生活水平，那么就绝对不可以忽视文化在发展中的重要性。我们也必须要更清楚地意识到，要改变这一点，不能继续我们现在的发展模式，不管是巴黎协定还是可持续性发展目标都提到，很多的国际政府和国际组织也做了很多努力来实现可持续发展。但是关于这一点，我认为理解和支持只是第一步，要真正达到这一目标，不仅仅是知晓，需要的是我们去改变态度和行为，转变其背后的激励和动机。经济发展、社会发展、环境保护对可持续发展都是非常重要的方面。研究显示，尽管目前我们对可持续发展有一个较为全面深入的理解，但是在态度和行为方面所做的改变，还远远不够。

所以接下来的问题是我们如何来鼓励这样的改变，我们怎么样让全社会意识到在我们生活中，文化对可持续发展的重要性。尊重文化是我们在进行可持续发展的过程中必须充分考虑的因素，我们必须要考虑到当地的文化，以及如何把不同群体的历史、文化和传统习俗融合。生活方式决定了个人行动和反映，要实现可持续发展的生活方式我们就必须要从改变做起。人类的发展也必须要通过行为和态度的变化来实现，比如在联合国发展署的一个在中国云南省开展的项目中，我们帮助少数民族妇女发展刺绣手工艺，把传统刺绣手工艺品销往本地乃至全国甚至国际市场，这样的项目不仅能帮助她们增收，同时可以有助于实现当地的可持续发展，还可以通过支持当地妇女让她们更好地继承文化和遗产，同时让年轻的一代参与进来。另外一个体现文化和传统知识的重要性的例子，是我们在海南省的一个关于农业生物多样性的工程，在这个叫作种子保护者的项目中，一部分人收集当地作物和当地其他植物的种子，存在种子银行当中，这样就可以帮助当地人们提前做好应对自然灾害的准备；同时我们把种子放在本地市场和其他地方的市场进行交易，不仅可以扩大种子银行的种类，而且能提高当地人对抗自然灾害的能力。保护本地的文化传统和知识要从保护其多样性开始，这样才能够实现我们整个生态文明建设的目标。

文化在之前的一些发展的论坛中并没有引起足够的重视。在发展的目标当中，有三个是关于文化重要性的，包括制定和施行政策推进可持续发展，宣传本地文化和产品、发展当地旅游业和创造就业机会，保护世界自然和文化遗产。可持续性发展和生态文明与我们的未来息息相关，我们必须更加重视我们的文化发展能力，并且通过我们自身努力改变我们的文化和行为。文化是我们生活中非常重要、不可忽视的方面，它不仅能帮助我们实现转型，而且能让我们的转型更顺利，实现可持续性发展的未来。

谢谢各位！

# Cultural Change for Sustainable Development

Ms. Agi Veres

## Personal resume

Ms. Agi Veres is the Country Director of the United Nations Development Programme (UNDP) in China.She is responsible for the overall management of UNDP's programming and operations in China, as well as partnership building and strategic coherence.

Ms. Agi Veres has been working for UNDP since 2002 in various capacities at Headquarters (HQ), regional and country levels. Prior to her appointment as Country Director in UNDP China, she was the Deputy Resident Representative of UNDP in Lesotho from October 2011. She worked at the regional level as Senior Programme Coordinator in the Bratislava Regional Centre for Europe and the Commonwealth of Independent States (CIS)(2009-2011), and Deputy Chief for Policy Support and Programme Development of the Bratislava Regional Centre (2006-2008). She started her UNDP career in Headquarters, including as Reporting Coordinator and Business Analysis

Team Lead (2005-2006), and ICT Communications Officer (2002-2005).

Ms. Agi Veres is a native of Hungary and has a degree of Master of Sciences in Business Administration from the Budapest University of Economic Sciences.

Thank you very much. Ladies and gentlemen, It is my pleasure to be here and address this very important event in which culture is firstly be a part of Guiyang Ecological forum. Ecological civilization is much related with sustainable development, they have much in common.

According to the ecological footprint as all of the time we refer to, the meaning issue is that if we keep continue our current way of living, if we keep buying or consuming the same way, our current population rolls rapidly, then, according to some studies, by 2030, we may need two earths to live in. But it is very unlikely that we will discover the second planet in the next 15 years. So to avoid this situation, our best bet is to change our culture of consumption and production. For the human development respective, that the UNDP think the overuse of our planet will be in bad for the poor and vulnerable people, and worse affecting how family security and develop issues. So from cultural respective, the topic I want to release today is what we need to suffer if we continue our current life, buying and consuming. The idea of what cultural and behavior change requires allow for more sustainable future and assure that our future generation could enjoy life without us jeopardizing the ability and have the same access to resources we live today. So when we look at the importance of cultural, let me quote by Amarty· Sam about relationship of culture ecology civilization, she said cultural matters are intangible parts of life living, if the development can be seen at the enhancement of improving our living standards, that efforts of culture in development can hardly be ignored. We should be more aware that to change and to achieve it we cannot continue our mode of development. Many countries and international communities have strived for sustainable development for years, as the Paris Agreement and Development Aim mentioned before. But actually to achieve this, it's the need to change our attitude, our behaviors and the conversion of its incentive and motivation behind. Economic and social development and environmental protection are very important for sustainable development. Though we have a deep knowledge of development, according to some survey, the change of our attitude and behaviors is still not enough, and much more level of change in attitude and behaviors should be done. So the next question is how can we encourage cultural change and how can we do this by also being in mind for the need for sustainability of cultural visible life.

Culture can be and must be considered in the sustainable development. We must consider local culture and the combination of narrative history and traditions that define their communities. Visual life make the people act and react. No transition could sustainable life style would be successful without taking parting change into account. For instance, one of the projects that the UNDP doing in China is a project in Yunnan province which help native women with their embroidery and help to sale the products to national and international markets. This is important for income generation and providing sustainable development for those villagers, and is equally important for local women's culture and heritage life and entire the younger generation get involved. Another project that underlines the importance of agriculture biodiversity in Hainan province is called seeds protecting project. The seeds collectors save the seeds of crops and plants against disasters. The more interesting part is that parts of the seeds from the seeds bank expanding the seeds bank. Maintaining local cultures, tradition and knowledge is not only about localization, because culture diversity of a homeland will make our homeland.

Ladies and gentlemen, Culture has not been paid enough attention to in the past tribunes. Targets for development include three about culture, they are implement policies to promote sustainable development, promotion of local cultural products, tourism and employment, protection of natural and cultural heritage. Because sustainable development and ecological civilization are all concerned about future rewards, we should pay more attention to the culture capacity to development, and through our own efforts to change our culture and behavior. Culture is not just an important fabric, it can help us not just do the transformation that is necessary, but was also possible. Thank you very much for your all attention!

# 生态文明贵阳国际论坛 2016 年年会
## Eco Forum Global Annual Conference Guiyang 2016

Speakers
演讲嘉宾

齐鸣秋

Qi Mingqiu

时任中国宋庆龄基金会常务副主席
Standing Vice-Chairman of China Soong Ching Ling
Foundation

　　齐鸣秋常务副主席有 10 年的国家宏观经济综合部门工作经验，18 年的国有大型企业领导经验，6 年的中央人民团体领导经验。获得工科和文科双学士学位、工程管理硕士学位，高级经济师、高级工程师。2016 年 5 月，齐鸣秋代表宋庆龄基金会与俄罗斯"文明对话"世界公众论坛签署合作备忘录，共同推动文明交流互鉴领域合作，被写入 2016《中俄联合声明》。

# 多元文化的和谐共生与新时代生态文明

齐鸣秋

时任中国宋庆龄基金会常务副主席齐鸣秋演讲

**尊敬的各位领导、嘉宾，女士们、先生们、朋友们：**

大家下午好！

首先，感谢组委会的邀请，荣幸地来到美丽的贵阳出席生态文明贵阳国际论坛。谨此向首度召开的"文化主题"论坛表示热烈祝贺，向为论坛举办做出积极贡献、付出辛勤劳动的朋友们表示崇高敬意。

习近平主席系列重要讲话多次指出，要把生态文明建设放在突出地位，融入经济建设、政治建设、文化建设、社会建设各方面和全过程。今年生态文明贵阳国际论坛首开"走向生态文明新时代的大文化行动"主题论坛，探讨如何将"文化"引入生态文明建设，推动文化和生态的深度融合与良性互动，这种探讨对于贯彻落实国家总体战略精神、建设新时代的生态文明具有重要意义。

生态文化是人类的文化积淀之一——由特定的民族或地区的生活方式、生产方式、宗教信仰、风俗习惯、伦理道德等文化因素构成的，具有独立特征的结构和功能的文化体系，是代代沿袭传承下来的针对生态资源进行合理摄取、利用和保护，促使人与自然和谐相处，可持续发展的知识和经验等。每个国家、民族和地区的生态文化都应该得到尊重、理解和包容。在此基础上的真诚对话、相互信任和友好合作是国际和平与安全的最佳保障之一。

在当今世界政治经济格局下，多元文化的共生共存已经成为一种必然现象。对此，中国传统文化提倡不同文明之间应该和谐共生。西周末年思想家、哲学家伯阳父认为，实现和谐则万物生长繁衍，如果完全一致，则无法发展延续。就像生物多样性是自然生态系统平衡的基石一样，多元文化的和谐共生是人类社会发展的根本。伯阳父之后，孔子提出"君子和而不同，小人同而不和"，将"和而不同"比作君子为人处世的基本原则。在当今的世界舞台上，中国始终倡导以海纳百川的胸襟，推动世界各国各民族不同文化的和谐共生。

中国宋庆龄基金会自 1982 年成立以来，一直致力于推动世界多元文化的交流，通过开展各种人文交流活动，增进不同国家、不同民族和不同文化的青年、妇女、少年儿童等群体的相互了解，进而推动多元文化的和谐共生。多年来，基金会作为活跃在国际舞台上的非政府组织，通过与全世界几十个国家和地区伙伴的紧密合作，开展了各种形式和主题的人文交流活动，取得了一定的成绩，赢得了国内和国际社会的积极评价。

2016 年 6 月下旬，俄罗斯总统普京访华，与习近平主席签署了中俄联合声明。其中第四项第 22 条提到，中国宋庆龄基金会和大型国际非政府组织——俄罗斯"文明对话"世界公众论坛于 2016 年 5 月在北京签署的合作备忘录，以及中俄两国非官方机构和其他国家相关组织共同倡议将丝绸之路作为连接沿线国家和人民传统的文化桥梁，这些都有助于促进社会组织在上述领域的合作，并于 2016 年 4 月在阿塞拜疆首都巴库召开的第七届联合国不同文明联盟全球论坛通过的决议中得到了体现。

以上内容是我会在中俄全面战略协作伙伴关系新阶段主动服务国家大局的具体体现。重点工作纳入《中俄联合声明》，是国家从促进中俄睦邻友好角度，对基金会多年来增进国际友好、维护世界和平事业给予的充分肯定，也激励我们继续加强与国内外相关机构合作，尊重文化的多样性和差异，推动不同文明的建设性协作，促进不同文明相互补充。

除了开展对俄的人文交流工作，中美、中欧、中印、中巴等大国关系及与周边国家及地区关系也是我会进行对外人文交流所关注的方向。近年来，基金会积极拓展与美国、欧洲、亚洲相关机构的合作，开展了很多学术及青年交流活动。美国方面，2015 年我会与美国耶鲁大学共同主办"和谐·合作·发展·责任——应对全球气候变化的理念与实践"国际圆桌会议，吸引了来自

10 个国家的 30 多位气候变化、哲学、法律、传播等领域的权威人士和知名学者参加。会议形成的成果呈报相关部委和中央领导后，得到较高评价。

欧洲方面，我会不断加强与德国知名学术机构墨卡托基金会、席勒研究所及法国戴高乐基金会等相关机构的联系与合作，双方不断互派学者参加对方主办的人文主题论坛，均取得了良好的效果。近期我会与匈牙利驻华使馆共同举办的中匈文化交流活动被列为外交部"2016 中国—中东欧国家人文交流年系列活动"之一。

亚洲方面，2015 年我会与印度世界事务委员会共同主办了"中印友好与文明互鉴圆桌会议"；基金会积极参与建设"一带一路"中巴经济走廊的进程，多次参加巴基斯坦驻华使馆的相关活动，目前正在策划开展中巴青年学生"一对一"交流项目；另外，我会还将于 7 月底在北京组织 2016 年"宋庆龄杯"中国东盟青少年足球友谊赛，邀请东盟 7 个国家参赛，该活动已被纳入外交部"中国—东盟建立对话关系 25 周年纪念活动清单"。

此外，作为具有联合国特别咨商地位的非政府组织，我会还积极加强与联合国相关机构的合作，2015 年我会与联合国开发计划署在北京联合主办了以"妇女·发展·未来"为主题的中外妇女论坛。

2016 年下半年，基金会还将组织一系列国际人文交流项目，比如为推动世界青年增进了解、凝聚共识、守护和平、共创未来，我会正与北京大学积极筹划举行"世界青年论坛"，届时将邀请世界知名大学的大学生、各行业的精英等约 300 名青年齐聚中国，就共同关心的话题进行交流，增进来自不同文化的青年之间的互学互鉴。

在基金会众多的国内外合作伙伴中，贵州始终是我们重要的站点之一。贵州是中国古人类发祥地之一，拥有悠久的历史，文化源远流长、意蕴深厚、特色鲜明、绚丽多彩。在贵州，多民族和谐共居，具有浓郁且神秘的民族风情，同时，贵州还有着独特的地理环境、舒适的气候条件，这些有利条件使得贵州在体现多元文化和谐共生特质的人文交流方面具有独特优势。多年来，贵州省委、省政府高度重视生态文明建设和国际人文交流，为推动中国优秀文化"走出去"和将外国朋友"请进来"开展人文交流做出了精心筹划，采取了切实行动，取得了积极成效。我一直在思考，如何在原有合作的基础上，加强对贵州的自然、人文资源的保护与利用，如何把更多的国际友人带到贵州，同时把多彩的贵州推向更高、更广的国际舞台。

近年来，我会已经在贵州开展了一些人文交流和公益项目，取得了较好的效果。例如，持续5年实施保护苗族传统手工艺项目，同时推进苗绣在保护传承中走向世界。2014年10月我会在法国巴黎举办的"中国苗绣展"，将贵州的苗绣推向巴黎时装周，该展览被列入国家"中法建交50周年重点活动"，受到中法各界关注，目前，我们正在积极运作该项目于2016年10月在美国联合国总部的展演活动。未来，我会将在人文交流、学术会议和艺术展演等领域与贵州展开多种层次及形式的合作。人文交流方面，针对我会的三大服务人群——青年、妇女、少年儿童，希望贵州能够成为上述三个群体进行国际交流的平台，让更多的国际友人，特别是外国青年大学生与中国大学生以"一对一"的形式走进贵州，了解贵州深厚的民族、民俗及民间文化，进而加深他们对中华文化的深入了解和对中国人民的友谊。学术会议方面，我会与中国社科院合作的"后海智库"将在生态文明与经济发展等领域，与贵州共同打造主题国际论坛，邀请世界各国的专家学者展开相关话题的深度探讨。在艺术展演方面，我会刚刚落成的中国宋庆龄科技文化中心可以作为贵州少数民族艺术传播、展演、展示和人才培养的基地，实现双方互补性合作。我们坚信，在贵州促进经济社会发展，特别是融入文化元素的生态文明新时代建设方面，我们有着巨大的合作空间和良好的合作前景。

女士们、先生们，朋友们！

今天，在中国与世界各国大发展战略的对接中，中国宋庆龄基金会作为推动多元文化繁荣共生、促进人类和平进步的一支力量，我们将发挥自身优势，在促进民间友好、民意沟通和民生合作中做得更多，做得更好。我们同样期待通过合作，同大家一起，弘扬生态文化，引领社会和民间机构积极行动，向世界展示中国人民热爱和平、友好环境、传承优秀文化、推动文明进步的美好愿景和积极实践。

最后，预祝大会圆满成功，祝愿贵州的生态文明建设事业蒸蒸日上，祝各位工作顺利，身体健康。

谢谢大家！

# Harmonious Coexistence of Multiculture and Ecological Civilization

Qi Mingqiu

## Personal resume

Qi Mingqiu has served in a comprehensive department in China's macro-economic administration for 10 years, in state-owned enterprises as a leader for 18 years and in a people's organization as a leader for 6 years. He holds double Bachelor's degree in engineering and arts, and Master's degree in engineering administration. He holds certificates of Senior Economist and Senior Engineer. Mr. Qi Mingqiu, on behalf of China Soong Ching Ling Foundation, signed the MOU between China Soong Ching Ling Foundation and Russian "Dialogue of Civilizations" World Public Forum in May, 2016, which has been written into *The Sino-Russian Joint Statement* in 2016.

## Respectable distinguished Guests, Ladies and Gentlemen,

Good afternoon! First of all, please allow me to express my gratitude to the organizing committee for their invitation to this grand event. I would also like to send my sincere congratulations to the opening of the cultural-themed sub forum for the first time. My respect also goes to those people who have made their contributions to the holding of the forum.

President Xi has attached great importance upon ecological civilization, which should be integrated into all aspects and overall process of economic, political, cultural, social construction as he has pointed out. Holding of the themed subforum on "Great Cultural Activism in the New Era of Ecological Civilization", which will focus on how to introduce "culture" into ecological civilization construction in order to promote better integration and interaction between culture and ecology, will be of great importance in carrying out our national strategy and construction of ecological civilization in the new era.

Eco-culture, as a part of human cultural accumulation, which is comprised of such cultural factors as specific ethnic or regional styles of life, production, religion, customs and ethics, is a cultural system with its own distinctive features and functions. It has been passed down over generations about how to take, use and protect ecological resources properly in order to promote harmonious coexistence between human beings and the nature and sustainable development. Eco-culture of every state, nation and region should be respected, understood and tolerated, which provides the basis for honest dialogue, mutual understanding and friendly cooperation in order to safeguard international peace and security.

Coexistence of multiple cultures is inevitable under the current international political and economic conditions, upon which the traditional Chinese culture advocates harmonious coexistence among different civilizations. Bo Yangfu, a thinker and philosopher at the end of the Western Zhou Dynasty, argued that all living things on earth will survive only when harmony is among diverse species achieved. On the contrary, nothing will survive if all the things enjoy no differences. In other words, coexistence of multiple cultures provides the basis for human development as what biodiversity is a footstone of balanced development of ecological system. Following Bo Yangfu, Confucius has taken "harmony in diversity" as the basic principle for people of moral integrity with his well-known saying "A gentleman gets along with others, but does not necessarily agree with them; while a person without moral integrity does the opposite." And in modern China, we always believe we should be tolerant enough to promote harmonious coexistence of multiple cultures.

Since its founding in 1982, China Soong Ching Ling Foundation (CSCLF) has devoted to carrying out various cultural people-to-people exchange programs in order to promote better understanding among youth, women and children from different parts of the world, which advances harmonious coexistence of multiple cultures. Being an active non-governmental organization in the international arena, CSCLF has achieved a lot in organizing people-to-people exchange programs with diversified themes and forms through close collaboration with its dozens of partners across the world, which has won high evaluation by both domestic and international communities.

China Soong Ching Ling Foundation has been written into *the Sino-Russian Joint Statement* signed by the Chinese President Xi Jinping and President Putin, his visiting Russian counterpart in late June this year. Article 22 reads like this: China Soong Ching Ling Foundation and Russia "Dialogue of Civilizations" World Forum, an NGO, endorsed a MOU in Beijing, May, 2016. Besides, other NGOs from both sides advocate taking Silk Road as a cultural bridge connecting the nations and peoples along the line. All of these joint activities will help accelerate cooperation between these NGOs in the mentioned areas and have embodied the resolution passed by the 7th UN Alliance of Civilizations International Forum held in Baku, the capital city of Azerbaijan, April of 2016.

The fact that the Foundation has been mentioned in *the Sino-Russian Joint Statement* shows that what the Foundation has done for world peace has received positive evaluation since the Foundation has made its working plan on the basis of serving our country in the new era of Sino-Russian Comprehensive Strategic Cooperative Partnership. And this fact will also encourage the Foundation to do better in promoting more constructive collaboration with its foreign partners in order to make more space for cultural diversities.

Besides Sino-Russian cultural people-to-people exchange programs, the Foundation has also devoted much to academic and youth exchanges with such other nations and regions as the United States, Europe, Asia and so on. In 2015, the Foundation collaborated with Yale University in holding the International Roundtable Conference on Climate Change with the theme of "Harmony, Cooperation, Development, Responsibility", in which more than 30 scholars from 10 countries took part. Its findings and achievements have been highly evaluated by the concerning government departments.

In Europe, the Foundation has also strengthened its cooperation with such prestigious institutes as Mercator Foundation (Germany), Schiller Institute(Germany) and De Gaulle Foundation(France). Recently, the Sino-Hungarian Cultural Event jointly held by the Foundation and the Hungarian Embassy in China has been put on the list of events for "China-Central and Eastern European Countries Cultural and People-to-People Exchange Year of 2016".

In Asia, the Foundation held "China-India Friendship and Civiliza-tional Exchange Roundtable Conference" with cooperation with India World Affairs Committee. The Foundation has played an active role in building Sino-Parkistan Economic Corridor under the frame of "One Belt, One Road" and taken part in the concerning events organized by the Pakistan Embassy in China. At present, a Sino-Pakistan Youth Peer-to-Peer Exchange program is being designed and planed. For the ASEAN countries, the Foundation will host 2016 "Soong Ching Ling" Youth Football Match, which has been included into the list of "Events to Commemorate the 25th Anniversary of the Establishment of Dialogical Relations Between China and ASEAN".

In addition, the Foundation has also kept close cooperation with the UN, as an NGO with Special Consultative Status with ECOSOC. An International Forum on Women collaborated with UNDP was held in Beijing last year.

For the second half of the year, a series of international cultural people-to-people exchange programs will be carried out among which the "World Youth Forum" which aims to promote better understanding and friendship among young people from the international communities.

Among the Foundation's multiple partners, Guizhou province has always been one of the most important ones. The Guizhou provincial government has achieved much in ecological civilization construction and international people-to-people exchanges through introducing Chinese culture into the international communities and inviting people from all over the world to China. In doing this, Guizhou enjoys its own advantages in terms of ecology and culture. On the one hand, the natural conditions have been well preserved here in Guizhou. On the other hand, Guizhou is the hub of colorful ethnic and non-material cultural heritages. These conditions give Guizhou unique advantages in promoting harmonious coexistence of multiple cultures and people-to-people exchanges. What I have been thinking about is how to have more foreign friends come to Guizhou and bring Guizhou into the international arena through cooperation between Guizhou and the Foundation and better preservation of the natural and cultural resources.

In this respect, we have achieved much in collaboration between

Guizhou and the Foundation in the areas of people-to-people exchanges and philanthropic programs. I would like to share a project with everyone here, which aims to preserve and publicize Miao Embroidery from Guizhou and has been carried out for 5 consecutive years. An Exhibit of Miao Embroidery was held in Paris, France during the Paris Fashion Week in October, 2014, which has been put on the list of "Key Events to Commemorate the 50th Anniversary of the Establishment of Diplomatic Relations Between China and France" and drawn great attention from both sides. At present, we have a further plan to move this project to the UN headquarter in the United States in October this year. In the future, we will conduct all-round cooperation with Guizhou in fields including culture exchange, public charities, academic conference, and art exhibition.

In the field of culture exchange, Guizhou will become an ideal platform of our international cooperation and exchange programs for our three focusing groups, which are youth, women, and children. And we hope more international friends, especially foreign youth will come to Guizhou in the form of "peer-to-peer" exchange accompanied with Chinese counterparts. By conducting such activities, we wish our international friends will know more about Guizhou people, its local folklore and its traditional culture, and then deepen their understanding about Chinese culture and promote the friendship with Chinese people.

In terms of academic conference, we will continue our cooperation with Chinese Academy of Social Sciences, and our Houhai Institute will focus on the field of ecological civilization and economic development. We will also work with Guizhou to set up international forums with different themes, in which we will invite scholars and professors all over the world to take in-depth discussion on related topic.

In the field of art exhibition, the newly-built China Soong Ching Ling Science and Culture Center will be our base for Guizhou ethnic minorities'art communication, art exhibition and personnel training in order to achieve complementary cooperation of the two parties. We firmly believe that we have huge potential and bright prospects for future cooperation with Guizhou, in the aspects of promoting Guizhou's socioeconomic development and developing the new era of ecological civilization with culture elements.

Ladies and gentlemen, dear friends, China Soong Ching Ling Foundation will spare no efforts in promote friendship, understanding and cooperation by making better use of our own advantages as a power to advance harmonious coexistence and world peace. At the same time, we hope to show the Chinese people's love for peace, friendliness, wish to inherit traditional culture and advance civilization to the whole world through our closer cooperation in carrying forward ecological culture, guiding actions by NGOs and the public.

To end my speech, I would like to wish a complete success of the forum and ecological civilization construction in Guizhou.

Thank you.

# 生态文明贵阳国际论坛 2016 年年会
## Eco Forum Global Annual Conference Guiyang 2016

Speakers
演讲嘉宾

梅里·玛达沙西

Mehri Madarshahi

全球文化网络主席，华南理工大学客座教授
President of Shenzhen−Qianhai MAH Global Cultural
Network Visiting Professor of Institute for Public Policy
(IPP).South China University for Technology, Guangzhou

　　梅里·玛达沙西曾作为高级经济官员在联合国任职 26
年（1977~2003 年），并且还于此期间被任命为联合国秘书长
权利下放特别工作小组成员，联合国改革委员会执行秘书高
级顾问等。1996~2009 年，她还曾是非洲领导基金会董事之
一。自 2004 年起，她出任文明间对话音乐协会和全球文化网
络的主席，并且获得包括阿斯彭文化外交奖（2012 年）在内
的多项国际奖项。并担任华南理工大学公共政策研究院客座
教授，深圳前海文化咨询公司首席执行官。

# 文化产业在经济增长和可持续发展中的角色

## 梅里·玛达沙西

玛达沙西女士出席论坛

感谢主席先生给我机会来谈谈文化在促进可持续发展问题中所起的重要作用。虽然我过去 8 年都参加了生态论坛会议，但很高兴主办单位第一次将这一重要议题列入会议议程，我在此表示祝贺。

最近一个由国际作曲者协会所做的研究显示，文化和创意产业对全球经济的贡献已经达到了 2.25 万亿美元，占全球经济总量的 3%。文化产业创造了 3000 万的工作岗位，这相当于全球劳动力总量的 1%。今天，文化和创意产业是两个被广泛使用的概念，被认为是各国经济的主要驱动力，它们为世界各国创造收入、提供就业、增加对外出口，在经济上被认为是一个越来越重要的角色。

文化一直以来被认为是促进社会和谐的有利因素。我相信文化也有语言，一种我们能够共通的语言，正如音乐能在全球共通一样。

通信服务在全球的盈利额为 1.5 万亿美元，而文化和创意产业在全球的收入已经超过了通信服务业，甚至超过了印度的国内生产总值（1.9 万亿美元）。文化和创意产业中盈利排名前三的行业分别是电视（4770 亿美元）、视觉艺术（3910 亿美元）、报纸杂志（3540 亿美元），名列雇用人数前三位的行业分别是视觉艺术（673 万人）、图书（367 万人）和音乐（398 万人）。

文化和创意的内容也促进了数字产品的销售，2013 年其销售额达到 5300 亿美元。数字文化产品目前是行业内最大的盈利来源，2013 年盈利 660 亿美元，其中在线媒体和免费视频网站的广告收入为 217 亿美元。表演艺术，如舞蹈、绘画、雕塑、公共事件的庆祝，目前在非正式经济中雇用了最多从业人员，他们为观众提供免费、非正式的音乐和剧场表演。2013 年非正式文化产业在新兴经济体国家（指发展中国家和欠发达国家）的产值达到 330 亿美元，提供了 200 万个工作岗位。文化产业日渐展现出其在促进经济增长和发展上的巨大潜力，然而它的潜力却没有得到充分发掘，常常在国家和国际发展政策和项目中被低估和边缘化。文化和发展之间联系密切，每个国家都应该制定高度相关战略和时政框架来认可文化的角色。

为此，国际社会一直在行动，过去十年有关文化和发展之间联系的概念性辩论既激烈又硕果累累，为建立与这方面相关的框架提供了战略和经验的基础：2004 年联合国发展署人类发展报告第一次较全面地阐述了文化自由对人类发展的重要性；在 2005 年联合国教科文组织发表的保护和促进文化表达多样性的公约中，有 80 个国家加入公约；2009 年在布鲁塞尔举行了以文化和创意为发展元素的学术研讨会。而且，联合国也推动达成了一系列协定，如千年发展目标（2000~2015），约翰内斯堡宣言，国际安居地联盟，2012 联合国可持续发展会议和可持续发展 2030 议程，2015 年提出的 17 个可持续发展目标，这些都极少提及文化在可持续发展中的重要作用，忽视了文化在创收和带动就业方面对经济增长的贡献。

越来越多的人意识到，要成功地解决全球性问题，如经济、发展、道德和经济增长等，我们需要新的思考方式、新的道德规范和价值标准以及新的行为模式。今天文化产业作为经济增长和出口的有力引擎，在创收和增加就业方面扮演的角色已经很难被忽视。文化产业对创新、出品、传播、增收、减少贫困、政治互信和出口具有战略意义。

一项可持续发展的战略，在文化上不可能是中性的。一种对文化差异敏感的发展方式，对化解经济、社会和环境问题至关重要。尽管有如此多成功案例和事实，但要引起经济学家对文化和可持续发展的足够关注和认可仍旧很困难。

尽管如此，我们知道让经济学家认同文化与可持续发展的关系是多么困难。所以让我暂时站在经济学家的角度说明问题。很少人知道，文化和创意产业是全球经济发展最快的领域之一，它们在非洲的平均增长率是 13.9%，在南美洲是 11.9%，在中东地区是 17.6%，在亚洲是 9.7%，在大洋洲是 6.9%，在北美洲和中美洲是 4.3%，这些增长率比前十年经济高速发展的中国还要高。我还想提一点就是，文化旅游占全球旅游总量的 40%，其对文化遗产及文化创造力的创意依赖性很高。

文化，包括创意产业，可持续旅游和遗产城市复兴都具有强大的经济带动力，可以创造绿色就业，促进地方发展和增加贸易机会，特别在发展中国家，他们更能依托自身丰富的文化遗产和具有创造力的人民来发展本国。

因此，这也让决策者更加强调相关文化部门经济增长和扶贫工作的贡献。特别是考虑到当前的经济危机，文化产业的影响需要被更好地理解。

公共政策应该成为文化、创意产业领域的核心，并支持创意产业以多种形式进行发展。理论框架必须转化为国家级的行动计划，进而带来对文化政策全面的、跨领域的关注，成为社会转型和保障公民权的重要工具。中国的社会科学和文化产业之间的关系还不够紧密，所以我觉得应该从国家层面发动整体性的行动计划，来加强文化和社会科学的联系，从而去实现我们的绿色经济可持续发展。因为文化因素会影响个人行为、环境管理的价值观以及与自然环境相处的模式，所以新政策应强调文化多样性在应对生态挑战、气候变化、防止生物多样性流失和确保环境的可持续性等方面的重要作用。

换言之，文化不仅可以带来经济上的盈利，还能改变观念、改变社会。文化和创意产业同时也可以创造非货币性的价值，可以实现以人为本的发展。所以在国家政策中，我们应该如何权衡和衡量文化和创意产业，这是一个值得深思的问题。如果一个国家有长期性的政策，其实施结果就应该可以将其转化为软性的基础设施，比如教育、科学、沟通、信息等。所以可持续发展只能依靠这种新型的模式，这是一种超越单纯追求经济盈利的发展模式，它会满足人类的需求和愿望，我想这也是我们所需要的。

谢谢大家！

# The Role of Cultural Industry in the Economic Growth and Sustainable Development

Mehri Madarshahi

## Personal resume

Mehri Madarshahi, served for 26 years (1977-2003) as a Senior Economic Officer at the United Nations Secretariat, she was assigned as a member of the UN Secretary-General's Task Force on the Decentralization and as a senior advisor to the Executive Secretary of the Reform Board of the UN. She also served as a director of Board of the Africa Leadership Foundation from 1996-2009.Since 2004, she acts as the President of "Melody for Dialogue among Civilization Association" and Global Cultural Networks and a recipient of many international awards including Aspen Diplomacy Awards in 2012. She is the Visiting Professor at the Institute for Public Policy of the South China University for Technology and the CEO of the Shenzhen-Qianhai Cultural Consulting Company (MAH). She is the Cultural Advisor to the Eoc Forum Global, and acts as the International Cultural Advisor to the Eoc Forum Global.

Good afternoon! Thanks Mr Chairman for giving me the opportunity to speak about the important role that culture plays in advancing sustainable development issues. I have attended the meetings of the Eco Forum in the past 8 years but I wish to congratulate the organizers for including this important topic in the agenda of the meetings for the first time.

As the latest study undertaken by the International Confederation of the Societies of Authors and Composers indicates, the contributions of cultural and creative industries to the world economy stands at US$2.250 trillion or 3% of the world GDP, creating some 30 million jobs worldwide equivalent to 1% of the world's active population. Today, the cultural and creative industries are considered as major drivers of national economies and in essence income generating, job creating and export earning factors in various countries of the world.

As it stands, the creative cultural industries revenues worldwide exceed those of telecom services (US$1,570b globally), and surpass India's GDP (US$1,900b). Within the total, the top three earners are television (US$477b), visual arts (US$391b), newspapers and magazines (US$354b). The top three employers among these industries are visual arts (6.73m), books (3.67m) and music (3.98m).

Cultural and creative content also helps sales of digital devices with a total of US$530b in 2013. Digital cultural goods are, by far, the biggest revenue source, generating US$66b in 2013 and US$21.7b of advertising revenues for online media and free streaming websites.

Performing arts are the biggest employers in the informal economy, providing unofficial music and theater performances, which are often free for audiences. Informal cultural industries sales in 2013 among emerging countries were estimated at US$33b providing 2 million jobs.

Despite the fact that cultural industries are becoming a dynamic sector with great potential for economic growth and economic development, their potential remains frequently underutilized and marginalized in national and international development policies and programs.

During the past decade, conceptual debates on linkages between culture and development have been both intense and fruitful allowing for the setting up of a highly relevant strategic and empirical framework: starting with the 2004 UNDP Human Development Report emphasizing the importance of cultural freedom for human development; to the 2005 UNESCO Convention on the Protection and Promotion of the Diversity of Cultural Expressions; the 2009 Colloquium on "Culture and Creativity as Factors of Development" in Brussels,. However, the major agreements reached at the United Nations such as MDGs (2000-2015), the Johannesburg Declaration, the Habitat International Coalition, the United Nations Conference on Sustainable Development in 2012 and the 2030 Agenda for Sustainable Development and its 17 SDGs in 2015 did marginally refer to the important role of culture in sustainable and economic development and undervalued the contribution of culture to economic growth in terms of both income and employment generation.

Ladies and Gentlemen,

There is a growing belief that to successfully advance in solving global problems, we need to develop new methods of thinking, to elaborate new moral and value criteria, and, with no doubt, a new pattern of behavior. Today, the role of cultural industries as a powerful engine for economic growth and export, generating considerable income and employment can hardly be ignored. Cultural industries are indeed a strategic outlet for innovation, production, dissemination, income generation, poverty reduction, political recognition and influence.

A sustainable development strategy cannot be culturally neutral. An approach to development sensitive to cultural differences is the key to addressing the interlinked economic, social and environmental problems confronting the planet as a whole.

Despite all, we know how difficult it is to rally economists to the culture /sustainable development nexus. Let me stay with the economists for a moment. Few know that cultural and creative industries represent one of the most rapidly expanding sectors in the global economy with an average growth rate of 13.9% in Africa, 11.9% in South America, 17.6% in the Middle East, 9.7% in Asia, 6.9% in Oceania, and 4.3% in North and Central America – these rates are by and large even higher than the often invoked growth rate of the dynamic Chinese economy in the past decade. Let me also mention that cultural tourism represents 40% of international tourism receipts and relies heavily on cultural heritage and creativity in all their forms.

Culture as well as creative industries, sustainable tourism, and heritage- based urban revitalization are powerful economic activities that generate green employment, stimulate local development and trade opportunities, particularly in developing countries with their often rich cultural heritage and creative population.

Thus it is left to the policy makers to highlight better the relevance of the culture sector's contribution to the economy and to poverty alleviation. Cultural industries' impact needs to be better understood and appreciated, especially in view of the current economic crisis.

Public policies should instill at the heart of the cultural and creative industry sectors support for the diverse forms of creativity. The framework must be translated into national action plans, which can bring a holistic and inter-sectorial focus to cultural policies as an instrument of social transformation and citizenship. The new policies should underline the important role of cultural diversity in tackling ecological challenges, coping with climate change, preventing biodiversity loss and ensuring environmental sustainability since cultural factors influence individual behavior, values related to environmental stewardship and the ways in which we interact with our natural environment. In other words, in addition to its economic benefits, the role of cultural and creative industries in generating non-monetary values that contribute significantly to achieving people-centered, inclusive and sustainable development must be elaborated upon.

Ladies and Gentlemen,

To this end, effective policies remain strategically important.

Success in attracting creative industries or creative actors can be seen as an outcome of long term policies affecting soft infrastructures, education, the sciences, culture and communication and information. Our sustainable future can be guaranteed only by a drive for a new type or form of development, one beyond the motivation of purely economic profitability. One, which by necessity drives to satisfy human needs and aspirations, as, declared to be the major objective of development.

Thank you for your attention.

# 生态文明贵阳国际论坛 2016 年年会
## Eco Forum Global Annual Conference Guiyang 2016

Speakers

演讲嘉宾

---

朝戈金

Chao Gejin

中国社会科学院学部委员、中国社会科学院民族文学研究所所长

Academician of the Academy of Social Sciences, researcher of National Literature Institute

---

　　朝戈金，法学博士。中国社会科学院学部委员、中国社会科学院民族文学研究所所长、研究员。曾在国外大学和科研机构中长期研修，有著作、论文数十种发表。主持或参与了联合国教科文组织和国务院系统几个部门的项目工作和发改委、文化部、中国文联等部委的委托项目工作。

# 国际非遗保护理念的解读：
# 我们在非遗保护中应当注意什么？

朝戈金

朝戈金研究员认真做笔记

大家下午好！我今天想讲的是伦理原则，我理解的生态包括自然环境的生态和社会文化的生态，在非遗保护过程中对伦理原则的遵守，是我们国家在这些年非遗工作中有所欠缺的。所以，我想特别就国际上关于非遗保护的伦理原则讲几点意见。

非物质文化遗产的保护原则，包括 2003 年的《保护非物质

文化遗产公约》和现有的保护人权和原住民公约等国际文件，这
是一套鼓励性原则。2015 年 11 月 30 日在非洲纳米比亚，召开
了联合国教科文组织保护非物质文化遗产政府间委员会第十次会
议，会议制定了 12 条保护非物质文化遗产伦理原则。这些原则
在今天和今后，对中国和国际社会保护非物质文化遗产具有非常
强的指导意义。在一定程度上匡正了学界和社会各界在非遗工作
中某些认识上的偏差。这 12 条原则，我在这里不逐条念，只谈
一些要点。

比如，相关社区群体和个人在保护其所持有和传承的非物
质文化遗产过程中应发挥主要作用。这句话听起来很普通，但是
事实上，我们回想一下，在政策制定、一些计划和设施落实的时
候，以遗产持有人与相关社区和民众为主的非遗保护工作，是不
是都做到了？我觉得不是都做得很好。

比如，社区群体与个人传承非物质文化遗产的权利，应得到
认可和尊重。就是说他们传或者不传是他们自己说了算，我们以
旁观者、局外人身份做出的某些推断，可能需要再斟酌。

比如，在非物质文化遗产保护的国际交流活动中，要特别强
调彼此的尊重和欣赏。我们生下来就在特定文化中成长，都带着
自己的文化眼镜，看其他文化的时候，有时会感到不解和困惑。
联合国教科文组织采取了一些举措来预防、消除这些疑惑。但像
杰出、特异、独一无二、原生等这样一些术语，我们不应将其过
于用于自己的文化，觉得自己的东西特别突出和独特。

比如，各类与非遗相关的活动要保证透明和开放，与遗产
持有人协商。我们现在有些基金、社会组织，扶持某些地方的非
物质文化遗产项目时，是不是应该在这件事上先让遗产持有人知

情，这方面的有些做法需要我们反思。即便在特殊情况下，如战争中，非遗持有人有权使用和利用与非物质文化遗产传承相关的场所、器具等。这些非遗传承的传统应该得到尊重，虽然这在一定程度上限制了传播。

比如，非遗价值的判断和认定应主要由社区民众和遗产持有人自己完成，不应当受到外界判断的影响。这是非遗和其他文化遗产不同的地方。人类文化遗产由专家组评定，非遗不是，这是非遗很不同的地方。

伦理原则不只是对外界有要求，对社区和遗产持有人也有特别要求。比如相关社区群体和个人有义务对其所传承的遗产可能造成的直接或者潜在威胁进行评估，他们有义务保护好、传承好非物质文化遗产。社区群体和个人在过度商业化、歪曲使用等方面也有责任防止或减少这些情况的发生。

文化多样化和文化认同要得到充分的尊重。性别平等、青年人的参与等因素在今后的非遗评定、非遗保护过程中，在条件允许、不违反民众意愿的情况下，应该被充分考虑。非物质文化遗产的保护是文化环境、文化生态保护相当重要的方面，对人类文明、人类和平、可持续发展具有十分重要的促进作用。

谢谢大家！

# Interpretation of Intangible Cultural Heritage Protection Concept: What should We Pay Attention in the Intangible Cultural Heritage Protection?

Chao Gejin

## Personal resume

Chao Ge Jin, doctor of law, academician of the Academy of Social Sciences, researcher of National literature Institute, who has been teaching in foreign universities and research institutions and published dozens of publications, papers. He chaired or participated in the projects supported by the United Nations Educational, scientific and Cultural Organization (UNESCO) and the State Council system several departments, the national development and Reform Commission, the Ministry of culture, the China Federation of literature and other ministries .

Good afternoon! The topic of my speech is ethical principle. The ecology I understand is the combination of natural environment and social culture. However, it may not be good enough for us to abide by the ethical principle in intangible cultural heritage protection. So I like to make a few comments on this issue here.

Principles of intangible cultural heritage protection include "non-material cultural heritage convention", a series of international documents on human rights, etc. The 10th intergovernmental meeting on intangible cultural heritage protection of UNESCO was held in Namibia of Africa on November 30, 2015, where 12 ethical principles were established, with great guiding significance in practice and correcting inaccurate perceptions of specialist and social friends in the intangible cultural heritage protection. Let's focus just on the leading points.

Firstly, individuals and community play a major role in intangible cultural heritage protection. All right, so now we ask ourselves, have we done well when we create and implement plans? I think the answer is NO.

Secondly, the individuals and community's rights of inheriting intangible cultural heritage should be recognized and respected. In other words, they can decide if they want to pass down the heritage, but we're kind of outsiders really.

Thirdly, we should respect and appreciate each other in the international exchange activities of intangible cultural heritage. Because of that there are different cultures in different countries or regions, we sometimes feel puzzled when we look at other culture with "colorblind glasses". To solve the problems, UNESCO takes measures for the prevention and elimination of people's confusion, for example, some terms like *outstanding, specific, unique* and *original* should not be overused to describe your own culture.

Fourthly, activities should be transparent and open through negotiating with the holders. That is, we should let holders know when some social groups give support to local projects of intangible cultural heritage. Even in the war, the tradition of inheritance should be respected, though there are limits to spread to some extent.

Fifthly, the judgment and identification of heritage value is mainly finished by the holder himself and community groups without external influences, while other heritages are indentified by expert groups, which is a typical characteristics of intangible cultural heritage.

However, the ethical principle is not only the judgment on the outside world, but also the requirement of communities and heritage holders. Community groups and individuals have the obligation to assess the immediate or potential threats to their heritage, and then protect well. In addition, community has responsibility to prevent or reduce over-commercialization and negative phenomenon in the intangible cultural heritage protection.

Culture diversity and identification should be well respected. Gender equality, participation of young people should be a consideration in the evaluation and protection of intangible cultural heritage without violating people's wishes. In short, interpretation of intangible cultural heritage protection focuses on the cultural environment and ecology, which plays an important role in human civilization, human peace and sustainable development.

Thank you!

# 生态文明贵阳国际论坛 2016 年年会
## Eco Forum Global Annual Conference Guiyang 2016

Speakers
演讲嘉宾

本·布尔

Bernhard Willem Boer

IUCN（世界自然保护联盟）环境法委员会副主席

The vice-president of Enviroment Law Committee of
Interational Union for Conservation of Nature

　　本·布尔是 IUCN（世界自然保护联盟）环境法委员会副主席、武汉大学环境法研究所"千人计划"特聘教授。他是武汉大学在文科领域引进的首个"千人计划"人才，也是世界闻名的顶尖环境法学者。现为武汉大学环境法研究所全职特聘教授，*Journal of Environmental Law*（SSCI 期刊）编委。他专攻环境法、比较环境法和国际环境法，著作丰硕。曾任 IUCN 环境法学院执行主席，现任 IUCN 全球环境法委员会副主席。2015 年世界自然保护联盟环境法学院授予其"资深学者奖"，以认可和表彰该学者在环境法领域的杰出学术贡献。

优秀文化引领生态文明新时代

# 中国生态文明文化与环境法治*

本·布尔

本·布尔教授认真听取发言

　　本文讨论了中国新兴的生态文明概念（或生态文明）被完全接受的过程，中国环境法的起草和实施方式需要从文化变迁方面理解。本文讨论了与文化相关的环境法起草和实施的一些方式。第一个是传统文化在促进生态文明以及早期的、有关可持续发展

　　* 本文为武汉大学本·布尔教授及同事所承担的教育部"生态文明与国际环境法"项目部分成果（No. 16JJD820012）。

概念方面的作用。第二个是生态文明在促进文化方面的作用,本文关注生态文明文化在促进中国环境法方面的问题。

生态文明理念被广为接受,环境法扮演了至关重要的角色,同时也为重新调整经济发展与环境保护的平衡做出了重要贡献。然而,现行有关环境保护的伦理和自然资源开发的法律需要改变。

首先,可以认为生态文明的概念大致等同于国际公认的"可持续发展"概念。自 1994 年《中国 21 世纪议程》出版以来,可持续发展一直是中国环境领域的热门词语。[1] 秦天宝教授指出:"2012 年,中国提出了'生态文明'概念,这被认为是中国可持续发展的表达方式,进一步推动可持续发展原则的发展。"[2] 亦可以进一步认为,生态文明包含了可持续发展的概念,可以被看作一个更深层次的概念,因为它超越了发展的可持续性的理念,需要我们研究人类与自然的关系的伦理基础和地球资源的利用。

回顾生态文明概念的发展历史,我们知道,其由中国前国家主席胡锦涛在 2007 年首次提出。[3] 这个概念在 2012 年中国共产党第十八届全国代表大会被列为"十三五"规划的五个目标之

---

[1] China's Agenda 21: White Paper on China's Population, Environment and Development in the 21st Century.

[2] Qin Tianbao, "China and international rule of law and environmental protection in ZENG," Lingliang and FENG Jiehan, eds., *Annual Report on China's Practice in Promoting the international Rule of Law*, Social Sciences Academic Press (China) 2015.

[3] "It is not a term the Party has coined just to fill a theoretical vacancy in its socialism with Chinese characteristics, but rather a future-oriented guiding principle based on the perception of the extremely high price we have paid for our economic miracle," *China Daily*, October 24, 2007, http://www.chinadaily.com.cn/opinion/2007-10/24/content_6201964.htm.

一。2013 年，习近平主席在党的十八届三中全会强调，中国将实现生态文明的改革，他的意思是协调经济发展和环境之间的矛盾。①

## 可持续发展的目标，生态文明和环境法治

理解生态文明与 2015 年联合国可持续发展目标之间的联系是非常重要的。联合国可持续发展目标是由联合国发起、历时两年达成的，其中包含 17 个可持续发展目标和 169 个具体目标，并出版了《改变我们的世界：一项新的全球行动议程和 2015 年的可持续发展目标》的文件成果。② 中国已承诺在实现可持续发展目标中发挥更加积极的作用。③

可持续发展的目标对国际环境政策框架是一个非常重要的补充，几乎涵盖所有环境与发展方面，包括水和环境卫生、气候变化、海洋和沿海地区、陆地生物多样性和土地退化等最重要的环境问题。本文主要讨论目标 16，其主要内容为："促进有利于可持续发展的和平和包容社会、为所有人提供诉诸司法的机会，在各层级建立有效、负责和包容的机构。"特别是其中的具体目标 3，关注"促进国际和国家层面的法治建设，确保所有人平等地获得司法救济"。这个目标是完全符合下文将要探讨的环境法治概念。这个目标给所有国家的环境法律框架和机构改

---

① Zhang Chun, "China's new blueprint for an ecological civilization," *The Diplomat,* September 30, 2015, http://thediplomat.com/2015/09/chinas-new-blueprint-for-an-ecological-civilization/.

② UNGA A/RES/70/1, https://sustainabledevelopment.un.org/post2015/transformingourworld.

③ "Chinese President Xi Jinping makes his first speech at UN on SDGs," http://english.cri.cn/12394/2015/09/27/4201s897716.htm.

革带来重大挑战。这些机构包括与环境和自然资源相关的部委、行政管理部门和法院。我们注意到，在中国，包括最高人民法院在内的各级法院已建立超过 500 个专门从事环境法的部门。①

中国的生态文明概念与《改变我们的世界》的主要元素以及可持续发展目标是一致的。这一点可以在习近平主席 2015 年联合国的演讲中得到充分体现："中国致力于结合国内中长期发展战略和 2030 年议程来发展生态文明。"为了保证实施效果，由 43 个政府部门组成的协调机制已建立。在宣传 2030 年议程，以动员国内资源，提高公众意识，并为创造良好的社会环境等方面做出最大努力。中国还将加强跨部门政策协调，以审查和修订相关法律法规为目标提供政策和立法保障。②

在"十三五"规划（2016–2020 年）的指导原则中包括了"生态文明"内容，整篇规划文本也多次引用此概念。"十三五"规划为促进环境法改革打下了坚实的基础，其对中国政策和法律的潜在影响甚至比可持续发展概念在其他国家中产生的影响更为深远。③各国如何通过法律框架实践这些概念也值得进一步研究。

目前，我们需要的是更深层次地思考实现生态文明概念的实

---

① See Pring, G. and Pring C., *Environmental Courts and Tribunals*, United Nations Environment Programme 2016.

② *Executive Summary of China's Actions on the Implementation of the 2030 Agenda For Sustainable Development* 2016 National Voluntary Reviews at the High–level Political Forum, https://sustainabledevelopment.un.org/member-states/china.

③ One such country is Australia, which has incorporated sustainable development and a range of associated principles in environmental laws at both national government (for example, Environment Protection and Biodiversity Conservation Act 199) and state government level (for example, Environmental Planning and Assessment Act 1979, New South Wales.

践和范畴。2017 年 3 月的一份《中国日报》提到，"生态文明的
关键原则包括需要尊重、保护和顺应自然，致力于资源保护、环
境修复与保护、资源回收、低碳与可持续发展"。[①] 可以看出，把
可持续发展作为生态文明的一部分很奇怪，因为可持续发展相当
于与其他元素一同被包含在这句话中。

## 生态文明背景下的环境法改革

如果生态文明的概念被世界各地更广泛采用，则需审视国际
环境法的各个方面，以确保建立更全面和一致的法律框架，制定
更强大的"绿色"的立法。这类研究可以而且应该通过国际环境
机构，如联合国环境规划署、混合政府 / 非政府组织、国际自然
保护联盟完成。[②]

在国家层面，所有在生态文明概念下的国家、公司将意味着
要重新审视环境和自然资源法律，以建议改革其目标、范畴、定
义和实现机制，判断他们能否实现生态文明的理念。

对于中国来说，2014 年的环境保护法修订中已经取得了改革
成果，将生态文明的概念作为其基本目标之一。

## 生态文明定义提炼

克劳斯·鲍斯曼提出了生态文明的广泛定义，他指出："我们
把自己作为一个相互联系的地球系统和社区的一部分，我们意识

---

① "Xi leads Ecological Civilization," *China Daily* , March 22, 2017, page 5.

② The World IUCN Commission on Environmental Law, the IUCN Environmental
Law Centre, and the member universities and research bodies of the IUCN
Academy of Environmental Law are well placed to carry out these tasks.

到我们对当代和子孙后代的责任。"他认为，为了创造实现生态文明的深刻变革，我们必须审视构建和支撑我们的社会核心结构和体系的基本假设和伦理价值观。

## 环境法治与生态文明

环境法治可理解为在地方、国家、区域和国际层面的环境决策中国际公认的"法治"。近年来，来自中国政府的政策文件都强调整个政府和中国社会的法治。① 正如上文提及的"十三五"规划（2016–2020 年）指导原则中的生态文明。第二章总体原则包括促进和协调经济建设、政治建设、文化建设、社会建设、生态文明建设。

世界自然保护联盟关于世界环境法治宣言强调，加强环境法治是实现高水平的环境保护的关键。宣言重申了一些广泛接受的原则，并记录几个与环境法治相关的新兴原则。其中部分原则要求我们对环境法的起草和实施做出一些根本性改变。宣言称，环境法治以一些关键治理要素为前提，它们包括，但不限于以下方面。

a. 制定、颁布和实施明确的、严格的、可行的、有效的法律、法规和政策，有效地通过公平和包容的过程管理来实现环境质量的最高标准。

b. 尊重人权，包括享有安全、清洁、健康和可持续的环境的权利。

---

① See for example, "Develop a law–based country, government and society" in Xi Jinping, *The Governance of China*, Foreign Languages Press 2014, pages 160 to 162.

c. 确保有效遵守法律、法规和政策的措施，包括充分的刑事、民事和行政执法、环境损害的责任以及及时、公正和独立的争端解决机制。

d. 平等获取信息的有效规则，公众参与决策，以及获得公正的权利。

e. 环境审计和报告，以及其他有效的问责、透明度、道德、廉政和反腐败机制。

f. 利用最有效的科学知识。

在中国，引入这些治理要素来进行法律改革需要大量的协商与沟通。然而，上面所阐述的概念显然与生态文明的理念是一致的。

### 中国保护区法律和生态文明案例

近段时间来，中国国家公园建设引起广泛关注。总体而言，关于自然保护的法律，尤其是国家公园和其他保护区，比其他环境领域的法律滞后。正如全国人民代表大会常委会在最近的一份声明中表示，中国一直致力于引入新国内法律，使保护区获得最佳的法律保护效果。中国政府的重点是在自然保护区成为中国国家公园体系的总体规划的一部分的背景下提倡生态文明的概念，这个规划将由中国政府在 2017 年制定。

### 结 论

为了生态文明文化被更广泛接受，并能够真正重新调整解决国民经济需求与加强环境保护之间的平衡，环境法和环境机构的各个方面都需要严格审视。我的论点是，生态文明的概念能不能充分实现，取决于整套环境法的研究和改革能否建立整体和一致的法律框架。考虑到上述问题，提出了对环境法治及其相关的原则。

总之，尽管中国的生态文明建设已经取得长足的发展，但需要在 2014 年环境保护法修订与"十三五"规划的基础上做出更多的努力。重要的是要确保"生态文明"的概念作为现代中国文化和社会的一部分，变得根深蒂固。但我认为，如果这些法制改革没有进行强力推动，生态文明建设目标可能会更晚才能实现。

Great cultural endeavours leading to a new era of Eco-civilization

# The Culture of Ecological Civilization and the Environmental Rule of Law in China[*]

Bernhard Willem Boer

## Personal resume

Bernhard willem boer is the vice-president of environment law committee of International Union for Conservation of Nature, distinguished professor of environment law research institute of Wuhan University through "thousand talents program". He is the first talent who was introduced through "thousand talents program" in humanity area in Wuhan university and the world's leading scholar of environmental law. At present, he is the distinguished and full-time professor in environment law research institute of Wuhan University and an editorial broad member of *Journal of Environmental Law*. In 2015, in order to approve and commend his outstanding academic contributions in the

---

[*] This contribution forms part of a project entitled 'Ecological Civilization and International Environmental Law' led by Professor Ben Boer and colleagues at Wuhan University, approved by the Ministry of Education in 2015 (No. 16JJD820012).

area of environmental law, environmental law academy of
IUCN awarded him the honor of " senior scholar".

This paper argues that for the emerging concept of ecological civilization (or eco-civilization) to be fully accepted, a cultural change is
required with respect to the way in which environmental law is drafted
and implemented in China. There are other ways in which culture is
relevant here. The first is the role of traditional culture in promoting
in eco-civilization and the older, linked concept of sustainable
development.[①] The second is the role of eco-civilization in promoting
culture; the very essence of civilization is indeed a high appreciation
of human culture. However, this paper is concerned with the promotion of a culture of eco-civilization within Chinese environmental law.

It is argued that environmental law has a vital role to play in
order for of ecological civilization to be more broadly accepted and
truly successful in terms of redressing the balance between national
economic demands and the need for enhanced environment protection.
However, for this to occur, there needs to be a change in the prevailing
legal culture concerning the ethics of environmental protection on the
one hand and the exploitation of natural resources on the other.

At the outset, we can recognise that ecological civilization can be
broadly equated with the concept of internationally accepted concept

---

[①] See for example Ben Boer, Culture, Rights and the Post–2015 Development
Agenda, in *Heritage, Culture and Rights: published in Challenging Legal
Discourses*, Andrea Durbach and Lucas Lixinski (eds), Hart, 2017: also
https://papers.ssrn.com/sol3/papers.cfm?abstract_id=2850201.

of "sustainable development". Sustainable development has of course been part of China's environmental lexicon since 1994, when *China's Agenda 21* was published.[①] As noted by Professor Qin Tianbao: "In 2012, China put forward the concept 'eco-civilization', which is regarded as the Chinese expression of sustainable development, further promoting the development of the principle of Sustainable Development."[②] One could argue further, that ecological civilisation certainly incorporates the concept of sustainable development, but can be seen as a deeper concept, because it goes beyond the idea of the sustainability of development activities, requiring us to examine the ethical basis of the human relationship to nature and our use of Earth's resources.

In considering the history of the development of the concept of eco-civilization, we can note that President Hu Jintao mentioned it first in 2007.[③] In 2012, the concept was listed as one of five goals in China's development plan during the 18th National Congress of the Chinese Communist Party. At the Third Plenary Session of the 18th

---

① China's Agenda 21: White Paper on China's Population, Environment and Development in the 21st Century.

② Qin Tianbao, 'China and international rule of law and environmental protec-tion' in ZENGLingliang and FENG Jiehan (eds) *Annual Report on China's Practice in Promoting the international Rule of Law*, Social Sciences Academic Press (China) 2015.

③ 'It is not a term the Party has coined just to fill a theoretical vacancy in its socialism with Chinese characteristics, but rather a future-oriented guiding principle based on the perception of the extremely high price we have paid for our economic miracle.' 'Ecological Civilization,' *China Daily*, 24 October 2007, at http://www.chinadaily.com.cn/opinion/2007-10/24/content_6201964. htm.

Central Committee in 2013, President Xi Jinping stressed that China would implement ecological civilization reforms, by which he meant "reforms to reconcile contradictions between economic development and the environment".[①]

The sustainable development goals, ecological civilization and the environmental rule of law

It is important to understand the link between eco-civilisation and the United Nations 2015 Sustainable Development Goals (SDGs). The SDGs were negotiated under the auspices of the United Nations over a period of two years up to 2015. They incorporate 17 goals, and 169 targets associated with those goals. They were published as part of the longer document entitled *Transforming our world: A New Agenda for Global Action and the 2015 Sustainable Development Goals.*[②] China has pledged to play an active part in the achievement of the SDGs.[③]

The Sustainable Development Goals are a very significant addition to the international framework of environmental policy. They cover almost all aspects of the field of environment and development. The most important environmental matters covered include water and sanitation, climate change, oceans and coasts, terrestrial biodiversity

---

① Zhang Chun, 'China's new blueprint for an ecological civilization' September 30, 2015, *The Diplomat*, http://thediplomat.com/2015/09/chinas-new-blueprint-for-an-ecological-civilization/.

② UNGA A/RES/70/1 https://sustainabledevelopment.un.org/post2015/transformingourworld.

③ "Chinese President Xi Jinping makes his first speech at UN on SDGs", http://English.cri.cn/12394/2015/09/27/4201s897716.htm.

and land degradation. With regard to the subject matter of this paper, Goal 16 is the most pertinent. The Goal is to: "Promote peaceful and inclusive societies for sustainable development, provide access to justice for all and build effective, accountable and inclusive institutions at all levels." In particular, Target 3 of Goal 16 focuses on the promotion of the "rule of law at the national and international levels and ensure equal access to justice for all". This Target is entirely consistent with the concept of the environmental rule of law that is discussed below. The Target presents significant challenges for all countries in terms of the review and reform of environmental legal frameworks and the institutions that support them. Those institutions include the relevant ministries and administrative departments concerning the environment and natural resources, as well as the courts. With respect to the courts, we can note that over 500 divisions of courts specialising in environmental law have been established at every level in China, including in the Supreme People's Court.[①]

China's eco-civilization concept is consistent with the main elements of "Transforming our World" and with the aims of the Sustainable Development Goals. This can be seen by the statement made at the United Nations by President Xi Jinping in 2015: "China has made great efforts in its implementation, linking the 2030 Agenda with domestic mid-and-long term development strategies. The domestic coordination mechanism for the implementation, comprised

---

① See Pring, G. and Pring C. *Environmental Courts and Tribunals*, United Nations Environment Programme 2016.

of 43 government departments, has been established to guarantee the implementation. Great efforts has been be made to publicize the 2030 Agenda nationwide in order to mobilize domestic resources, raise public awareness, and creating favorable social environment for the implementation. China will also strengthen inter-sector policy coordination, review and revise relevant laws and regulations to provide policy and legislative guarantee[s] for the implementation."①

In the Thirteenth Five Year Plan 2016-2020, we see that the plan includes "ecological civilization" among its Guiding Principles, and has a number of references to the concept throughout the text. The Plan forms a solid foundation for promoting the reform of environmental law. It has the potential to influence Chinese law and policy even more fundamentally than the concept of sustainable development in other countries.② A comparison between the with regard to how the concepts are being implemented in practice through legal frameworks in each country would be useful avenue for further research.

For the moment, what is needed is deeper thinking about the scope of and practical implementation of the concept of ecological

---

① *Executive Summary of China's Actions on the Implementation of the 2030 Agenda For Sustainable Development* 2016 National Voluntary Reviews at the High-level Political Forum, https://sustainabledevelopment.un.org/member-states/china.

② One such country is Australia, which has incorporated sustainable development and a range of associated principles in environmental laws at both national government (for example, Environment Protection and Biodiversity Conservation Act 199) and state government level (for example, Environmental Planning and Assessment Act 1979, New South Wales.

civilization. According to a report in the *China Daily* of March 2017, "the key tenets of ecological civilization include the need to respect, protect and adapt to nature; a commitment to resource conservation; environmental restoration and protection; recycling; low carbon use; and sustainable development".[1] We can note here that the mention of sustainable development as part of this list is curious, as sustainable development can be seen to already incorporate the other elements included in this quote.

## The reform of environmental law in the context of ecological civilization

If the concept of ecological civilization were to be more widely adopted around the world, it would demand a critical examination of all aspects of international environmental law to ensure that more holistic integrated and consistent legal frameworks are generated with the view of drafting much more robust "green" legislation. Such research can and should be done through the work of international environmental institutions, such as the United Nations Environment Programme and the hybrid governmental/ non-governmental organization, the International Union for Conservation of Nature.[2]

At a national level, for all countries, the incorporation of the

---

[1] 'Xi leads ecological civilization' in *China Daily* , March 22, 2017, page 5.

[2] The World IUCN Commission on Environmental Law, the IUCN Environmental Law Centre, and the member universities and research bodies of the IUCN Academy of Environmental Law are well placed to carry out these tasks.

concepts underlying ecological civilization would mean taking a close look at environmental and natural resources laws in order to recommend reform of their objectives, scope, definitions, and implementation mechanisms in order to judge whether they are capable of achieving the ideals of ecological civilization.

For China, it already means building on the reforms already achieved in the 2014 Environment Protection Law, which incorporated the concept of ecological civilization as one of its basic objectives.

### Refining the definition of ecological civilization

Klaus Bosselmann has proposed a broad definition of eecological civilization as referring to "a world in which we govern themselves as part of an interconnected Earth system and community, mindful of our obligations to both the current generation and future generations". He argues that in order to create the profound transformations to achieve ecological civilization, we must examine the fundamental assumptions and ethical values that frame and underpin our societies' core structures and systems.

### The environmental rule of law and eco-civilization

The "environmental rule of law" is understood as the application of the internationally accepted "rule of law" at local, national, regional and international levels in the context of environmental decision-making. In recent years, the policy documents emanating from the Chinese government have emphasized implementation of the rule of law throughout the government and Chinese society in general.[1] In examining China's Thirteenth Five Year Plan 2016-2020, as noted, ecological civilization is included among its Guiding Principles. Chapter 2 includes the overarching guiding principle: "Promote and coordinate the economic construction, political construction, cultural construction, social construction, and ecological civilization construction."

The IUCN World Declaration on the Environmental Rule of Law[2] states that "Strengthening the environmental rule of law is the key to achieving the highest possible level of environmental conservation and protection." The declaration restates a number of broadly accepted principles, and records several emerging principles associated with the environmental rule of law. Several of these principles demand some fundamental changes in in the way we think about environmental law is drafted and implemented. The Declaration states:

The environmental rule of law is premised on key governance elements including, but not limited to:

Development, enactment, and implementation of clear, strict, enforceable, and effective laws, regulations, and policies that are efficiently administered through fair and inclusive processes to achieve the highest standards of environmental quality;

Respect for human rights, including the right to a safe, clean,

---

[1] See for example, 'Develop a law-based country, government and society' in Xi Jinping, *The Governance of China*, Foreign Languages Press 2014, pages 160 to 162.

[2] IUCN World Declaration on the environmental Rule of Law, adopted in April 2016 in Rio de Janeiro at the IUCN 1st World Congress on Environmental Law, available at http://www.unep.org/environmentalgovernance/erl/.

healthy, and sustainable environment;

Measures to ensure effective compliance with laws, regulations, and policies, including adequate criminal, civil, and administrative enforcement, liability for environmental damage, and mechanisms for timely, impartial, and independent dispute resolution;

Effective rules on equal access to information, public participation in decision-making, and access to justice;

Environmental auditing and reporting, together with other effective accountability, transparency, ethics, integrity and anti-corruption mechanisms; and use of best-available scientific knowledge.

In China, legal reforms to introduce these governance elements will require a good deal negotiation and drafting. However, the concepts spelled out above, are clearly consistent with the idea of eco-civilization.

**Example of China's protected areas law and ecological civilization**
In recent times, there has been a great deal of focus on the development of new national parks in China. The nation's laws concerning nature conservation in general, and national parks and other protected areas in particular, are less developed than other areas of the environmental realm.[1] China has been investigating optimal legal approaches

---

[1] 'As part of China's efforts to improve the environment, the construction of nine national parks with a combined area of nearly 170,000 square kilometers is underway, and the country will this year formulate an overall plan for the national park system': 'Xi leads ecological civilization', *China Daily*, March 22, 2017, at http://www.chinadaily.com.cn/cndy/2017-03/22/content_28633954.htm.

to protected areas with the view to introducing new domestic laws, as indicated by a recent statement by the Chair of the National People's Congress Standing Committee.[①] The focus of the Chinese Government is on promoting the concept of ecological civilization in the context of nature conservation as part of an overall plan for China's National Park system. This plan is intended to be formulated by the Chinese Government in the course of 2017.

## Conclusion

In order for the culture of ecological civilization to be more broadly accepted and truly successful in terms of redressing the balance between national economic demands and the need for enhanced environment protection, a critical examination of all aspects environmental law and environmental institutions is required. My contention is that the idea of the ecological civilization cannot be adequately implemented unless the whole suite of environmental laws is examined and reformed to ensure that more holistic integrated and consistent legal frameworks are generated, taking into account the issues that have been raised above with respect to the environmental rule of law and its associated principles.

---

① See Zhang Dejiang, Chair of NPC Standing Committee: 'With a continued commitment to green development and ecological progress, we will revise the Law on the Prevention and Control of Water Pollution, the Law on the Protection of the Marine Environment, and the Wildlife Protection Law, in a bid to put in place the most rigorous possible system for ecological conservation and environmental protection.' From 'Report on the Work of the Standing Committee of the National People's Congress', in China's NPC Approves 13th Five-Year Plan (Beijing: National People's Congress, 2016), 13.

In summary, although some excellent progress has already been made in China, much more effort is required by legal researchers and legal drafters to build on the reforms achieved in the 2014 Environment Protection Law and the relevant elements of the Thirteenth Five Year Plan. It is important to ensure that the concept of ecological civilization becomes deeply embedded, as part of modern Chinese culture and society. If such legal reforms are not strongly promoted, the aims of eco-civilization will be much slower to attain.

**生态文明贵阳国际论坛 2016 年年会**
Eco Forum Global Annual Conference Guiyang 2016

Speakers
演讲嘉宾

---

陈　田
Chen Tian

中国科学院地理科学与资源研究所研究员
Doctoral supervisor of Institute Geographic Science and
National Resources Research

---

　　陈田，中国科学院地理科学与资源研究所博士生导师、
旅游与社会文化地理研究室主任。先后主持和参与国家攻关、
院重大、基金和有关部委科研项目 30 多项。参与编著 10 多
部，发表论文 60 多篇。曾获部委科技进步奖一等奖二次、二
等奖两次，自然科学奖三等奖一次。

# 如何为地域文化繁荣培育蓬松和肥沃的土壤

陈 田

陈田研究员演讲

　　首先很高兴给我这样一个机会，让我讲生态文明，因为我有地理学背景，今天讲的题目是如何为地域文化繁荣培育蓬松和肥沃的土壤。

　　地域文化概念是人文地理里面独特的概念，一般指特定区域源远流长、独具特色、传承至今仍发挥作用的文化传统，是特定区域生态、民俗、传统、习惯等文明的表现。它在一定的地域范围内与环境相融合，因而被打上了地域的烙印，具有独特性。地域文化中的"地域"，是文化形成的地理背景，范围可大可小。地域文化中的"文化"，可以是单要素的，也可以是多要素的。地域文化的形成是一个长期的过程，地域文化是不断发展、变化的，但在一定阶段具有相对的稳定性。它的特征可以体现在很多方面，如方言文化、饮食文化、民间信仰、民间建筑、环境不同、移民影响、区划影响等，使得我们这个地域更加丰富多彩。

地域文化与生态文明的关系，我认为体现在三个方面：其一，生态文明，包含我们对自然生态和文化生态（人文生态）的认知、态度、价值观及制度安排；其二，地域文化客观上成为承载生态文明的基石和载体；其三，关注与发展地域文化，是践行生态文明理念的抓手，也是展示生态文明进步的窗口。

传统地域文化面临什么问题？我认为面临两大挑战。一是地域文化的形成需要依赖时间的积累和沉淀；而城市化、现代化、信息化与流动性过程加快，正在改变着孕育地域文化的土壤环境，包括养分和结构，导致地域文化特色遭受非理性的侵蚀而褪色。二是文化传播的碎片化、快餐化以及外来文化的侵蚀，地域空间破碎化和地域文化趋同。地域文化保护、传承与创新，需要我们有更大的智慧和魄力，也需要我们自问：百年之后，现存的地域文化遗产会怎样？我们又会给下一个百年留下怎样的遗产？

怎样肥沃地域文化繁荣的土壤？我有三点建议。一是树立全民敬畏心，让保护和繁荣地域文化真正成为决策管理者的常态意识。二是重视发展民间文化社团和文化社区，蓬松和肥沃地域文化发育的土壤。三是政府鼓励与支持，让地域文化的守护者、创造者能够在公益化和市场化浪潮中，获得持续、有益的养分，获得尊严，体现价值认同。谢谢大家！

# How to Create Fertile Soil for Regional Culture Prosperity

Chen Tian

## Personal resume

Chen Tian, doctoral supervisor of Institute Geographic Sciences and Natural Resources Research, director of Tourism and Social Cultural Geographic Research Office. He has taken charge of and participated in more than 30 programs ranged from ministry to state level. Besides, he has compile over 10 books and published over 60 papers. He has been awarded the first prize of scientific progress at the level of ministries and commissions for twice, the second prize for twice and the third prize of natural science for once.

First of all, I am very happy for having such a chance to talk about the ecological civilization. I majored in the geography, and the topic I tell is "How to Create Fertile Soil for Regional Culture Prosperity"?

Regional culture, as an important category of human geography, is an unique cultural tradition in certain areas with a long history, now it is still playing an important role. It is the sign of civilization of ecology, folklore, tradition and custom, having distinctive regional features. Region, the geographical background of the culture formation, may be large or small. Culture, can be a single factor or a multiple one. The regional culture is continuously developing and changing in the long-term formation, but shows a relative stability during a certain stage. Characteristics of regional culture has been mostly incarnated the respects of dialect culture, food culture, folk beliefs, folk architecture, different environment, immigration, district division and so on, which makes life more colorful in a certain region.

The relations between regional culture and ecological civilization are mainly reflected in three aspects: ecological civilization contains the cognition, attitude, value and institutional arrangement of natural ecology, cultural ecology (human ecology); regional culture, as the fusion zone of natural ecology and cultural ecology, is the cornerstone and carrier of ecological civilization; developing the regional culture to practice the ecological civilization idea and show its progress.

What are problems traditional regional cultures facing? I think we have two major challenges. First, it takes time for regional culture formation, while the process of urbanization, modernization and information is accelerated which leads to cultural erosion by regional environment changes. Second, the fragmentation of cultural transmission, fast food culture, foreign culture erosion, geographical space fragmentation and regional culture convergence are other challenges we are facing. So we need to have more wisdom and courage to do the work of regional culture protection, inheritance and innovation. The main thing for us is we need to analyze our position and ask ourselves: what is the existing regional cultural heritage? And after a hundred years, what legacy will we have?

How to create fertile soil for regional culture prosperity? I think there are three important points. Firstly, it is necessary for people to regard it with awe, for administrator to establish the sense of regional culture protection and development. Secondly, we should attaches importance to developing folk culture association and cultural community. Thirdly, government encouragement and support is of importance to inspire creativity and energy of the guardian and creator of regional culture, which can embody man's identification of value. Thank you!

# 生态文明贵阳国际论坛 2016 年年会
## Eco Forum Global Annual Conference Guiyang 2016

Speakers

演讲嘉宾

全京秀

Chun Kyung Soo

韩国首尔大学名誉教授、著名生态人类学家

Famous ecological anthropologists, honorary professor of
Seoul University of South Korea

全京秀，著名生态人类学家，韩国首尔大学名誉教授，
韩国人类学学会会长，2014 年韩国总统勋章获得者。发表论
文 300 多篇，著述 30 多部，其代表作为《环境人类学》。因
《粪便是资源》一书，其被称为"粪便博士"。

# 生态与文化：人类学视域下的贵州

全京秀

全京秀教授演讲

　　各位尊敬的贵宾，下午好！我是一位人类学学者。两年前在韩国退休，之后被邀请到贵州大学任教。在这段时间里，我开始了解贵州。今天，我非常高兴来到这里，不仅是作为演讲者，有机会向各位分享自己于生态和人类学方面的观察与思考；而且很高兴作为听众，有机会听到很多富有见解的演讲。

　　刚接触到生态文化这一领域时，我去参加了一个生态实验，当时对 DNA、RNA 都不算了解，后来通过研习生态方向的著作，我受益匪浅。自从来到贵州之后，我经常去贵州不同的地区去参观和考察当地人的生活现状和生态现状。

生态系统分为宏观和微观两个方面。《生态文明》这本书认为宏观层面上的生态系统包括社会阶层、等级等。如果把文化包括进来，人类社会和自然生态系统就有了区别，人类社会在一些植物、动物、农作物种植等方面会有自己的实践，文化对生态系统起到了非常大的影响。

先看微观生态系统。黔灵山位于贵阳市中心，在这座有 400 万人口的城市的中心生活着超过 5000 只猴子。我经常来观察猴子的日常生活，包括观察猴子的社会结构，发现猴子与猴子之间、猴子与人之间进行交流互动的方式和人类社会有很多相似性，比如它们的群体有不同的等级，要给猴王贡献食物，通过决斗决定群体内的地位。我觉得像猴子这样一个群体，其实是微观层面的生态系统写照。

再看宏观生态系统。在不同的村落里，自然资源、自然环境都为人类所用，如用木头建房子、用自然材料织网。我从农业领域学、人类学、生物多样性等方面进行了研究。一些农居、灌溉系统、厕所、养殖物，成为一个地区独有的生态文化和生态系统。一些地区少数民族开展祖先祭祀活动，宰猪、宰羊、杀牛，将其作为祭祀品，通过这样的方式和祖先对话，这是一种宏观生态系统。但是现在，基础设施、大坝、工厂进入这些村庄，对当地环境造成了污染。于是我理解了为什么在"十三五"规划中，会把生态文明建设作为重要目标。因为工业化从西方国家发达地区逐渐转移到偏远山区，给当地生态环境造成了影响。

现在再来看微生态系统。微生态系统，如食品的发酵，茅台酒就是细菌发酵的结果，贵州的很多茶叶是发酵茶；再比如疾病也是微生态系统的一部分。几千年前的孔子创立了儒教，他曾经说，抓鸟不能用笼子抓，因为一个笼子是有限的空间，抓了一次以后就很难再抓，也抓不到更多，这是一种历史的智慧，这种理念应该得到更好的发展。

中草药在贵阳也有很大的市场，贵州有苗族的中草药，侗族的中草药，种类丰富，如何把信息收集起来，对人类学家来说，这是很重要的问题。太阳是人类能量的来源，中国古代有后羿射日的传说，黔东南的妇女将太阳做成图案表现在她们的服饰上。将来我们可以探索更多有关生态文化和人类学的概念，这些是我们未来生活和文化的一部分。非常感谢！

# Ecology and Culture :
# Guizhou in the Anthropological Perspective

Chun Kyung Soo

## Personal resume

Chun Kyung Soo is a famous ecological anthropologists, honorary professor of Seoul University of South Korea, president of the Institute of Anthropology of Korea, winner of 2014 South Korean Presidential Medal, published more than 300 papers, 30 books. His representative work is *Environmental Anthropology*. Because *Feces is Resource*, he was called as "stool doctor".

Good afternoon, distinguished guests. I am an anthropologist. I retired in Korea two years ago, then I was invited to teach in Guizhou University. During this time I started to get an acquaintance of Guizhou province. I am very happy to be here today not only as a speaker but also as a listener to have the chance hear so much insightful thoughts and ideas.

When I entered into this area I attended an ecological experiment, I didn't know what DNA or RNA were. Reading and research of ecological books helped me lot. I often went to different places in Guizhou to observe and study their life and ecological conditions.

Ecosystem includes macro and micro ones. *Ecological civilization* thinks that macro ecosystem includes social ranks etc. If culture was included, then human society and nature ecosystem make a difference. Human's society practice on plants, corps and animals, and culture works a lot on ecological system.

Let's firstly talk about the micro ecosystem. In the picture we can see the Qianling mountain which located in central city with a population of four million people has over 5 thousand monkeys. I often observe their life and social structure. I found they have much in common with human beings in communication. They have ranks; they fight for status, and contribute food to the monkey king. I think the group of monkeys is an embodiment of micro ecosystem.

Let's then talk about the macro ecosystem. Natural resources and environment can be used in building wood houses and making fabrics. I made studies in agriculture, anthropology and biodiversity. Some houses, irrigation systems, toilets, domestic animals make the unique ecosystem. Some ethnic areas make sacrifice to communicate with ancestors by killing pigs, sheep, cattle. It's about the macro ecosystem. But today, infrastructures, dams, factures pollutes villages after they entered village. So I understand why ecological civilization was taken as an important goal in the 13th Five-Year Plan , because industrialization has been transferred from western countries to the remote areas, which is harmful for local ecological environment.

Now let's see micro system. Food fermentation, like Maotai spirit, a outcome of germ fermentation inside, is a micro system. Disease is also a micro system. Fermented tea is the major type in Guizhou. Confucius who established Confucianism thousands of years age once said that we couldn't catch birds into cages for the room is limited and it's hard to catch them again or to catch more. It is wisdom of history which should be well developed.

Chinese herbal medicine sells well and are of different kinds in Guizhou, like Miao and Dong herbal medicines. How to collect that information is a problem for us anthropologists. Sun is source of energy of human beings and there's a famous tale about sun and his 9 sons. In southeastern Guizhou, women make patterns of sun on their clothes. In the future we shall explore more concepts about ecological civilization and anthropology, and they are part of our future life and culture. Thank you very much.

# 生态文明贵阳国际论坛 2016 年年会
## Eco Forum Global Annual Conference Guiyang 2016

### Speakers
### 演讲嘉宾

卜希霆

Bu Xiting

中国传媒大学文化发展研究院教授

Professor of the Institute of Cultural Development, Communication University of China

 

 

卜希霆，中国传媒大学经管学部党委副书记、文化发展研究院副院长，副研究员，文化部公共文化研究基地主任。主要研究方向为公共文化、文化创意产业、创意营造学等。

# 生态城市与创意营造

卜希霆

卜希霆教授演讲

## 从"空气罐头"说起

今天生态文明贵阳国际论坛的举办地——贵州，被誉为"国家公园省"，空气质量备受关注，2014 年全国"两会"期间，中共中央总书记、国家主席习近平参加贵州团审议时称，"PM2.5 空气质量直接关系百姓的幸福感，将来贵州可以卖'空气罐头'"。

虽然是一句戏言，却也表达了习近平总书记对大自然的敬畏和对好生态、好环境的期待及对贵州好山好水的赞赏，亦传达了敬畏自然、爱护环境、保护生态的发展理念。

进入 21 世纪，随着城镇化的快速推进，城市变得日益多姿多彩；它带给我们的烦恼也与日俱增。喜剧电影《煎饼侠》的推广曲《五环之歌》一经上线便刷爆朋友圈并引发网友的学唱热潮，成为当下大热的洗脑神曲。虽然戏谑，但反映了"摊大饼"式的城市发展带给人们的尴尬，"五环"似乎成为城市发展的一个诅咒。人与城市的纠结与矛盾日益困扰着我们。

纠结之一：交通系统越来越发达，每个人每天乘坐交通工具上班的时间越来越长。更多的年轻人虽然留在大城市，但每天疲于漫长而憋屈的公交通勤；更有无数家庭为了孩子上学蜗居"学区房"，甚至不断改变着交通与出行方式。

纠结之二：钢筋水泥建筑越来越多，城市人的个体社会交往空间越发逼仄。人与人摩肩接踵于钢筋水泥的城市，舒适的交往空间却日益狭窄。自我空间稍被人触犯就会感到不舒服、不安全，甚至恼怒。更可怕的是，今天的年轻人干脆沉迷于虚拟空间，懒得面对真实世界。

纠结之三：信息系统越来越复杂和先进，反常识的"知识型文盲"越来越多。当下时代，知识更新加速，科技与时俱进，海量信息令人目不暇接，同时，群体盲目性抉择也让时代发展受到威胁。如何理性面对信息时代是摆在今天的时代难题。

纠结之四：个体的疾病想象与社会的疾病体质，造成城市人群的病态消费。今天城市人群的"社会疾病体质"造成的病态消费令人咋舌，甚至有人开玩笑讲现在最贵的车是"马云的购物车"……

交通堵塞、资源短缺、拥挤不堪、人口膨胀、环境污染、高昂房价……面对日益严重的"城市病"，越来越多的人开始厌倦城市，甚至"逃离城市"。城市呈现前所未有的时代危机：文脉断裂、千城一面、产城不容、矛盾加剧……面对狂飙式的城市化带来的"城市病"，我们必须适时反思，及时调整，从而转向城市内涵的提升、人居环境的改善、绿色生态的重建。

## 生态城市的重生之路

城市发展的历史告诉我们：作为一个城市，钢筋水泥并没有生命，一个恢弘的城市轮廓可以在短期内建造起来，但是一个生态文明的城市绝不会在短期内显现。生态城市的建设不应局限于自然的生态环境的保持，生态城市的可持续发展更是源自一种向上的人文力量。现代生态都市建设首先要构建良好的人文观念，再通过自然山水的修复，治理体系的构建，创意氛围的营造，这样才能改变水泥丛林的城市发展噩梦。

第一，人文养成。城市最好的模式是关心人和陶冶人，城市是人类赖以生存和发展的重要介质，它不仅是居住生息、工作、购物的地方，更是"文化容器"。生态城市的建设，首先要有一个共同的文化认知，应遵循"知行合一"的理念。"知行合一"对贵州有特殊的意义，不仅因为这个思想是王阳明先生在500年前的贵州"悟"出来的，更因为今天的贵州特别需要"知行合一"这样一种精神状态和行为准则。每个民族的文明都有自己内在的肌理，强行打乱必然造成血脉的闭塞、身体的萎痹。从国家发展而言，传统文化是治国理政的宝贵资源、内在血脉。优秀文化的涵养将有利于重构生态文化、培育生态伦理、增进生态共识。因此生态城市首先要从人文建设开始，更多吸取人文的力量，让城市走得更远、更长，有可持续性发展的动力。

第二，山水修复。生态城市是重构与自然平衡的城市，对城市已有生态系统进行自然化改造。通过对接近自然的森林与草地、湿地的共同发展与保护，建设和培育城市稳定高效的近自然生态系统。

重构自然平衡关系，是今天许多城市正在进行的努力，从钢筋水泥灰颜色的城市，慢慢开始有颜色，慢慢开始恢复到富有生机和绿色。通过已有生态系统的自然化改造，通过对接近自然的森林、草地、湿地的共同发展与保护，建设和培育城市稳定高效的近自然生态系统。近年来，生态文明贵阳国际论坛顺应世界生态文明建设的趋势和潮流，体现了中国优秀传统文化的智慧和情怀，展现了中国推动绿色发展、建设生态文明的务实行动，具有示范意义。

第三，科学治理。当今世界，科技进步日新月异，互联网、云计算、大数据等现代信息技术深刻改变着人类的思维、生产、生活和学习方式。科技创新将有助于经济、自然、文化等方面的快速发展。今天科技创新已经成为引领未来经济增长的主要动力，因此要推进城市管理机构改革，既要通过科技创新来进一步推进城市治理体系和治理能力现代化，又要通过充分挖掘大数据提升政府治理能力，提升服务民生水平，建立基于"智慧城市"的现代化城市治理模式，这也正是生态城市的题中之义。

第四，创意营造。一座城市能够被人们关注、具有吸引力，必然有它最为独特的一面。如果北京没有故宫、上海没有外滩、杭州没有西湖，她们将失去独特的魅力。在巴黎，埃菲尔铁塔、巴黎圣母院、卢浮宫构成法国历史的缩影；在迪拜，帆船酒店是国家的象征；在悉尼，歌剧院成为澳大利亚的标志……

创意是城市的灯塔，创造更美好的城市不仅是设计者的事情，而且是人人参与的创意过程。营造富于创意与想象力的城市环境氛围，是培育生态城市可持续发展的内生动力。创意生态的发展也离不开人的创造。创意生态要调动起大众的积极参与，创造宽松的文化氛围和强大的包容性，实现人的创意价值，使人人享有创意带来的福祉。通过创意生态的营造，使人的创意得到进一步实现，促进新文化、新思想的交融。

生态城市不会一蹴而就，需要细水长流；创意营造不在一朝一夕，需要坚持不懈。生态城市建设应以人文为魂、以自然为本、以科技为翼、以创意为先，生态城市在创意营造过程中，必须尊重传统、强调个性，并从城市发展文脉中寻求创意灵感。人文建设是未来生态城市发展的灵魂，而文化产业必然是生态城市的主导产业。创意环境是创意者创造的舞台，也是创意产生的客观背景，优质的创意环境可以激发城市的创意活力，进而推动生态城市的可持续发展。

# Eco-City and Creativity Establishment

## Bu Xiting

### Personal resume

BU Xiting, Deputy Dean of cultural development institute, School of Economics and Management, Communication University of China.

Bu Xiting, deputy party secretary, cultural development, Deputy Dean of the institute, associate professor, director of public cultural research base of Ministry of Culture. Research interest: public culture, cultural creative industries, creative learning, etc.

### From Canned Air

The venue of Ecological Global, Guiyang, Guizhou is honored as a province of parks, in which the air quality is under great concern. CPC central committee general secretary and state President Xi Jinping said in 2014:" PM2.5 air quality is directly related to the happiness of our people, Guizhou can sell canned air in the future". This word reveals Xi Jinping's respect to the nature and expectation to good environment, as well as appreciation of good preservation of landscapes in Guizhou. Furthermore, his word expressed belief of reverence for nature, care and protect the environment.

Entering into 21 century, with fast urbanization, more and more troubles appear. *The Song of Fifth Ring* from comedy movie *The Pancake Man* became popularized in We Chat when it appeared in the movie. Apparently, it was joked but mirrored the embarrassing situation we are facing in big cities. "The fifth ring" became a curse of city development. The embarrassing situation and dilemmas of man and cities are increasing.

Dilemma 1, urban transportation systems are becoming more and more developed, while the time for people to travel to work is getting longer and longer. More and more youngster stay in big cities, but are tired of long time and nurse agrievance trip to work. Numerous families have to stay Dwelling Narrowness in school districts for their children, or change their ways of transportation.

Dilemma 2, there are more and more reinforced concrete buildings, while the individual and social communication space of people in the city is narrower and narrower. People crowded in the concrete cities, but comfortable space has been narrowed day by day. Uncomfortable unsafe even irritable emotions are easily aroused by invasion of private space. More seriously, young people today are addicted to virtual space and unwilling to face the real world.

Dilemma 3, information systems are becoming more complex and advanced, while there are more and more intellectuals of anti-common sense. The speed-up knowledge update and technology, mass information make people unprepared. The group blind choice threats the development. Rational dealing with the information becomes a challenge.

Dilemma 4, individual imagination of diseases and social disease system lead to the morbid consumption of city people. The morbid consumption made by social disease constitution of urban people is astonishing. Some people joke that the most expensive car is Ma Yun's shopping cart, which is indeed a blind consumption.

The increasingly worsen urban disease such as traffic jam, resources shortage, population explosion, environmental pollution and high house prices make people be tired of cities or even escape from cities. The cities tend to appear culture fracture, similar cities, sharpen contradiction etc, which reflect the crisis of urbanization. These urban diseases should be adjusted and turn to increase the connotative capacity of the cities by improving living environment and rebuilding ecology.

### The way for city rebirth

How do cities get rebirth? The history of urban development shows that a magnificent urban contour with inanimate reinforced concrete can be built in a short period of time, but the eco-city cannot. The construct of eco-city is not limit to environment protection, but originated from humanity force to sustainable development. The construct of modern eco cities should be based on humanity belief, then through landscape recovery, governance system construction and creativity establishment to get rid of the nightmare of concrete forest cities.

First, humanistic spirit cultivation. The best mode of a city is to care people and cultivate people. City is the base for living, and not only a place for living, work and shopping, but a "container of culture". The construction of eco-city should be based a common cultural cognition, the belief of "unity of knowledge and action". "Unity of knowledge and action" was created by Wang Yangming in Guizhou 500 years ago, which means great for Guizhou. Guizhou needs these spirits and behavior standards. The culture of each ethnic group has its interior culture texture, the violence to break the rules may result in decline of the carrier. From the point of country development, traditional culture is the precious resources to govern the country. Splendid culture will favor to reconstruct eco culture, cultivate ecological ethic and strengthen ecological consensus. So, the construct of eco city take priority to humanity construction, therefore, the city will have a sustainable development with dynamic power.

Second, landscape rehabilitation. Eco city is to reconstruct the balance between city and nature. Through natural rebuild of existing eco-system, through conservation and development of forest, grassland and wetland, we could build and cultivate stability and high-performance close to natural eco-systems of cities.

Now, many cities are making efforts to reconstruct the balance of nature and city. They color the city with green from grey cement, recovery the vitality. Through natural rebuild of existing eco-system, through conservation and development of forest, grassland and wetland, we could build and cultivate stability and high-performance close to natural eco-systems of cities. The Eco Forum Global held in Guiyang these years complies with wave of world eco development, reflects the wisdom and feelings of Chinese excellent culture, shows Chinese practice achievements in pushing green development and eco construction, which sets an example for the world.

Third, scientific governance. In today's world, rapid progress in science technology and the Internet, cloud computing, big data and other modern information technology is profoundly changing the human way of thinking, production, living and learning. Science and technology innovation would help the rapid development of the economy, natural, cultural and other aspects. Today, science and technology innovation has become main engine of the future economic growth, so, we need to promote the reform of city governance institutions, as well as promoting the modernization of urban management system and management ability through technological innovation; increasing the service level of the people's livelihood and government management ability through the large data. The modern urban governance mode based on "wisdom city" is one characteristic of the ecological city.

Fourth, creativity establishment. A city attracted by people must have its unique characteristics. Beijing without the Forbidden City, Shanghai without the Bund and Hangzhou without the West lake will lost their charm. In Paris, the Eiffel Tower, the Notre Dame DE Paris, the Louvre constitute the epitome of the history of France; in Dubai, sailing hotel is a national symbol; the Sydney opera house becomes a symbol of Australia.

Creativity is the beacon of the city. Building a better city is not just a designer's responsibility, but a creative process for everyone to participate. Building the urban environment atmosphere of creativity and imagination is endogenous power for the sustainable development of ecological city. The creation idea of ecological development is inseparable from people. Creative ecology needs to stimulate the public to participate in creating the loose culture atmosphere actively, and then to realize people's creative value, make everyone enjoy the creative of well-being. Through the idea of ecological construction, we can realize the creativity; promote the union of the new culture and new ideas.

However, this process takes time and perseverance. Eco-city construction should stick to work under the guidance of "humanity crucial, nature first, science and technology important and creativity priority". Humanity construction is the soul of the future ecological city development, and the cultural industry is inevitably the leading industry of ecological city. Creative environment is the stage of creatives, as well as the objective background of idea generation. The high quality creative environment can stimulate the city creative vitality, and then promote the sustainable development of ecological city.

# 生态文明贵阳国际论坛 **2016** 年年会
## Eco Forum Global Annual Conference Guiyang 2016

Speakers

演讲嘉宾

徐　静

Xu Jing

贵州省文化厅厅长

Chief of Cultural Department in Guizhou province

　　徐静同志曾担任过贵州省社会科学院副院长、贵州省社会科学界联合会副主席，长期从事经济社会发展研究和党的理论政策研究及宣传工作。主持国家级和省级课题 30 余项，出版著作 20 余部，发表理论文章 180 余篇。是国家"万人计划"首批入选者，全国文化名家暨中宣部"四个一批"人才，享受国务院特殊津贴专家。

# 走向生态文明新时代的文化坚守

徐　静

贵州省文化厅厅长徐静演讲

大家好！今天听了雒部长的演讲，以及各位嘉宾的演讲，对我的触动很大。部长的演讲明确提出生态文化是生态文明建设的重要基石，同时还深刻阐述到：由于生态环境保护的复杂性、长期性，解决这个问题不仅要从体制、技术层面进行深入思考，更要从理念、文化等方面入手，树立尊重自然、顺应自然、保护自然的生态文明理念，形成天人合一、天人互益的生态文化，从根本上为生态环境保护的长期开展奠定基础。部长对文化在生态文明新时代中的地位和作用的论述，给文化工作者以巨大的信心和鼓舞，同时也指明了生态文明新时代文化工作的前进方向。作为一名贵州的文化工作者，大力弘扬优秀的传统生态文化，推动贵州生态文明建设是我们义不容辞的使命。

我们要大力弘扬"天人合一"的自然观。雒树刚部长的演讲指出，生态文化是生态环境保护的重要内核，是生态文明建设的重要基石，这要求我们把生态文明建设使命内化为建设者的价值追求，外化为建设者的自觉行动。历史上，贵州各族群众尊重自然、敬畏自然、顺应自然，留下了许多符合生态规律和生态价值要求的贵州经验、贵州故事，历史地积淀了"天人合一"的自然观。在走向生态文明新时代的今天，我们要更加自觉地挖掘、保护、传承和弘扬积淀在贵州地域上的优秀生态文化传统，增强当代人的生态意识、生态自觉和生态自信。

我们要大力弘扬"知行合一"的实践观。雒部长在演讲中指出，文化在推动生态文明理念形成和普及方面有启迪和引导作用，在"行"这一方面也能够发挥巨大作用。从部长的演讲中我体会到，大力弘扬生态文化，也要坚持"知行合一"，坚持把理念与行动的交融作为生态文明新时代的内在特质。贵州是阳明文化的发源地，王阳明被贬贵州后，受到纯朴、善良的贵州人民的关心和爱护，参悟出"知行合一"的实践观。因此，在生态文明既是理念又是行动的今天，我们更应该大力弘扬贵州历史上积淀的"知行合一"的实践观，推动生态文明新时代的文化建设。

我们要大力弘扬"美美与共"的和谐观。有一种观点认为，生态文明必须取代工业文明，建设生态文明就是要迈开工业文明阶段，这显然陷入了唯阶段论的困境。事实上，人类社会从来就是多元文明共存、共进、共荣的社会。历史上，工业文明的出现并没有取代农业文明，相反提升了农业的文明程度。今天提出生态文明，并不是要替代以前的工业文明、农业文明，而是要进一步提升我们工业和农业的文明程度。部长在演讲中指出，要以文化之力推动生态文明建设，要用文化提升我们的绿色空间、绿色产业、绿色生活方式，我的理解就是要用文化来"美人之美"，最终形成"美美与共"的局面。而这，也正是宋庆龄基金会齐鸣秋主席所倡导的多元文化的和谐共生的新时代生态文明的理想。在走向生态文明新时代的今天，我们需要更加包容不同时代、不同地域的文明，大力弘扬贵州历史上积淀的不同文明与文化"各美其美、美美与共"的和谐观，更加推动生态文明新时代经济社会的包容性发展，让生态文明的成果惠及全体人民。

最后，我想表达两个意思。一个就是对我们在座的各位表示感谢，我作为承办方衷心地感谢大家，感谢文化部，感谢宋庆龄基金会，感谢各位嘉宾的支持。另一个就是对我们在座的各位表示歉意，请求大家原谅，因为时间紧、任务重，论坛有很多不尽意、不到位的地方。请大家相信我们，我们在明年时候的相聚，无论是学术探讨水平，还是服务水平，我相信都会上一个大台阶。再次向大家表示深深的歉意和衷心地感谢，谢谢！

# Culture Persistence of Ecological Civilization in the New Era

### Xu Jing

## Personal resume

Xujing used to be the vice executor of Guizhou Academy of Social Science and vice president of Guizhou Federation of Social Science. She is occupied in long-term research on the development of economic society and research on theoretical policy and publicity of Chinese Communist Party. She has presided about 30 subsjects of national and provincial level , published nearly 20 works and about 180 theoretical articles. She is the expert for special allowance of the State Council.

Good afternoon, everyone! The Speeches of the minister of China Culture Division Luo Shugang and other distinguished guests touch me greatly. Minister Luo explicitly raised ecological culture is an important foundation of ecological civilization construction, and emphasized that we should not only start from the system, but also take into account the culture because of the complexity and long-term characteristic of environmental protection. Therefore, to achieve ecological civilization needs: to build the idea of complying with nature, respecting nature and protecting ecology, so as to form the ecological culture with the theory of "harmony between man and nature". The minister's statement on position and role of culture in the era of ecological civilization will give cultural workers great confidence and inspiration as well as point out the direction of the work in this field. As cultural workers in Guizhou, it is our responsibility and the sacred missions to carry forward the fine traditional culture of the Chinese nation and promote the construction of ecological civilization in Guizhou.

We should vigorously promote natural view of "harmony between man and nature". The minister said that ecological culture is an important part of ecological environment protection and ecological civilization construction, which require us to internalize value of seeking for ecological civilization construction and externalize voluntary motions of builders. In history, people of all nationalities in Guizhou respected nature, complied with the nature, left many valuable experiences and stories in accordance with ecological law and ecological requirement.

Nowadays, we should unearth, protect, inherit and promote the excellent traditional culture in Guizhou, for strengthening consciousness and confidence of people in the construction of ecological civilization.

We should vigorously promote the practical view of "the unity of knowledge and action". The minister said that: culture, as a guide, has important enlightenment for formation and popularization of ecological civilization idea. I learned that we should adhere to "the unity of knowledge and action" in vigorously promoting ecological culture. Guizhou was the cradle of Wang Yang-mings' culture. Wang Yangming, a philosopher in Ming Dynasty, was transferred to Guizhou as punishment because he was against imperial officials. The love and care of people in Guizhou gave him the momentum to create "the unity of knowledge and action". So we should vigorously carry forward the practice view of "the unity of knowledge and action", promoting the development of ecological culture in the new age.

We should vigorously promote the harmonious view of "co-existence with difference". However, an irrational argument in favor of that ecological civilization must replace the industrial civilization. In fact, the common progress by embracing diversity of civilization runs through many areas of human societies. And the industrial civilization promotes the agricultural civilization degree rather than replace it. Similarly, the ecological civilization will promote the development of the agriculture and industrial civilization. The minister said that we should promote the green space, green industry and green life through the culture, while the vice-chairman of China's Soong

Ching Ling Foundation Qi Mingqiu advocates the ideal of harmonious development and coexistence of diverse culture. In one word, it is the harmonious view of "co-existence with difference". We should vigorously develop different civilization and culture in Guihzou to promote the harmonious view of "co-existence with difference", so as to improve the inclusive development of ecological civilization and bring its benefits to all the people.

At last, I want to thank the support of the Ministry of Culture, China's Soong Ching Ling Foundation and all the guests who are participating. And we will apologize to you, if there is inconvenience during this session. We will make an effort to improve our service next time. Thank you very much!

# ◇ 多彩文化　美轮美奂 ◇

## Colorful Culture, Magnificent Province

　　贵州是中国唯一没有平原支撑的山地省份，是一片充满神奇魅力的文化沃土。在这里，生态多样性和文化多样性交融，青山绿水、古树名木、村落遗存、传统技艺、古老传说等特色文化元素，构成一幅多彩的图画、一个和谐的世界、一方心灵的港湾，寄托了无尽的乡愁。

Guizhou boasts its picturesque landscapes, pleasant climate and various and rich tourism resources. This is a magic and charming land famous for its unique culture. Here, biodiversity fuses with culture diversity, such as green hills and clear water, historical trees and famous wood species, village heritage, traditional skills, ancient legend, etc. Those special cultural elements forming together lay the frameworks of a colorful painting, a harmonious world, a harbor of hearts where left endless homesickness.

贵州侗族大歌
Guizhou Grand Choirs of Dong

传统村落文化遗产
Cultural Heritage of Traditional Villages

苗族姊妹节
Sister's Meal Festival of Miao

苗族芒筒芦笙
Mangtong Lusheng of Miao

苗族鼓藏节
Guzang Festival of Miao

# 亚鲁王
## The Yalu King

傩戏
Nuo Drama

风格迥异的苗族芦笙舞
Lusheng Dance of Miao

布依族八音坐唱
Eight Tones of Buyei

安顺地戏
Tunpu Dixi Opera

海龙屯土司遗址
Dragon Tuen Chieftain Site

苗族服饰
The Clothing of Miao

水族马尾绣
The Horsetail-based Embroidery of Shui

苗绣
Embroidery of Miao

彝族撮泰吉
Cuotaiji Show of Yi

牙舟陶器炼制技艺
Yazhou Pottery Craft

侗族琵琶歌
Pipa Songs of Dong

苗族蜡染技艺
Batik Craft of Miao

皮纸制作技艺
Bark Paper Manufacturing Craft

银饰锻制技艺
Silver Jewelry Making Craft

# ◇ 多彩文化　共建共享 ◇

## Colorful Culture, Co-construction and Sharing

　　贵州是舞的世界、歌的海洋，各族群众在这片土地上创造了多姿多彩的文化艺术。当前，贵州正深入挖掘特色文化资源，着力打造"多彩贵州文化艺术节"这一艺术的盛会、人民的节日，营造各族群众参与文化创造、分享文化成果的良好氛围，彰显贵州的文化自信。

Guizhou is a place where multi-ethnic people accumulate and create various and colorful cultures and arts. Currently, Guizhou aims to deeply explore its special culture resources and focus on constructing "the festival of cultures and arts of colorful Guizhou", a great event of art and of people's festival, and create a relaxing atmosphere of participating in cultural creation and sharing cultural achievements, which manifests the confidence of Guizhou culture.

歌舞剧《撮泰吉》
Music Drama: *Cuotaiji* (Yi language, Games that ghosts bless their descendants)

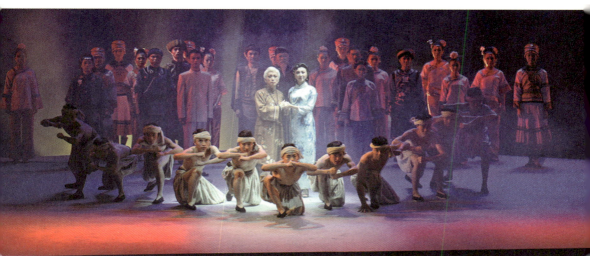

花灯剧《盐道》
Huadeng Opera: *Salt Transporting Road*

音乐剧《嘎老》
Music Drama: *AI Laox*

侗族歌舞剧《丹砂恋》
Music of Dong: *Love of Cinnabar*

多彩贵州文化艺术节
Colorful Guizhou Culture & Art Festival

《多彩贵州风》在世界各地演出
*Colorful Guizhou* was performed around the world

京剧《铁弓缘》
Chinese Opera: *Fate Comes From Iron Arch*

话剧《图云关》
Stage Play: *Tuyun Pass*

大型舞剧《天蝉地傩》
Large-Scale Musical: *Tian Chan Di Nuo*

莫斯科文化演出交流活动
Cultural Communicaticn in Moscow

舞蹈诗《巫卡调恰》
Dancing Poem: *Wuka Diaoqia*（Miao Language, meas grandma's ballabs）

黔剧《折子戏》
Guizhou Opera: *Opera of Highlights*

**马德里演出**
The Culture Communication in Madrid

阳明文化传习活动
Inheritance of Culture of The Yangming

长征文化传承活动
Inheritance of Culture of The Long March

非遗周末聚
Exhibition of Intangible Cultural Heritage

图书在版编目（CIP）数据

走向生态文明新时代的大文化行动：2016 生态文明
贵阳国际论坛生态文化主题论坛讲演集／徐静主编. --
北京：社会科学文献出版社，2017.6
ISBN 978 - 7 - 5201 - 0645 - 0

Ⅰ.①走… Ⅱ.①徐… Ⅲ.①生态环境建设 - 国际学
术会议 - 文集 Ⅳ.①X171.4 - 53

中国版本图书馆 CIP 数据核字（2017）第 074865 号

走向生态文明新时代的大文化行动
——2016 生态文明贵阳国际论坛生态文化主题论坛讲演集

主　　编／徐　静

出 版 人／谢寿光
项目统筹／王　绯　周　琼
责任编辑／周　琼　刘晶晶

出　　版／社会科学文献出版社·社会政法分社（010）59367156
　　　　　　地址：北京市北三环中路甲 29 号院华龙大厦　邮编：100029
　　　　　　网址：www.ssap.com.cn
发　　行／市场营销中心（010）59367081　59367018
印　　装／北京盛通印刷股份有限公司

规　　格／开 本：787mm×1092mm　1/16
　　　　　　印 张：11　字 数：123 千字
版　　次／2017 年 6 月第 1 版　2017 年 6 月第 1 次印刷
书　　号／ISBN 978 - 7 - 5201 - 0645 - 0
定　　价／69.00 元

本书如有印装质量问题，请与读者服务中心（010 - 59367028）联系